Industrial Electrical Wiring: Design and Application

John T. Earl

Prentice-Hall, Inc., Englewood Cliffs, NJ 07632

Library of Congress Cataloging-in-Publication Data

Earl, John T. (date)
Industrial electrical wiring.

Includes index.
1. Factories—Electric equipment—Installation.
2. Electric wiring, Interior. 3. Electric engineering
—Insurance requirements. I. Title.
TK3283.E19 1987 621.319'24 86-18642
ISBN 0-13-459520-3

Editorial/production supervision
and interior design: Sophie Papanikolaou
Cover design: Lundgren Graphics
Manufacturing buyer: Gordon Osbourne

Dedication

This book is respectfully dedicated to
AUGUST NOLLERT
who was the long-time plant manager of the
Virginia Oak Tannery in Luray, VA
and
who contributed much to completing my education
on industrial electrical wiring some years ago

Printed in the United States of America
10 9 8 7 6 5 4 3 2

ISBN 0-13-459520-3 025

Prentice-Hall International (UK) Limited, *London*
Prentice-Hall of Australia Pty. Limited, *Sydney*
Prentice-Hall Canada Inc., *Toronto*
Prentice-Hall Hispanoamericana, S.A., *Mexico*
Prentice-Hall of India Private Limited, *New Delhi*
Prentice-Hall of Japan, Inc., *Tokyo*
Prentice-Hall of Southeast Asia Pte. Ltd., *Singapore*
Editora Prentice-Hall do Brasil, Ltda., *Rio de Janeiro*

Contents

Preface

Texts on industrial electrical wiring have normally included page after page of theory, mathematical calculations, and equations with very little material covering practical applications. *Industrial Electrical Wiring: Design and Application* is designed to change this; that is, with the scores of texts available covering theory, this author chose to concentrate mainly on the practical applications of industrial wiring, leaving the theory portion to other quite qualified books on the subject.

This book begins with an overview of the industrial electrical industry, touches briefly on essentials such as codes, standards, and blueprint reading, then quickly jumps into practical categories such as tools, electrical materials, equipment, and the design and implementation of actual installations.

The plant engineers and electricians, obviously, cannot afford to be without this book, but others in the electrical industry will also benefit: consulting engineering firms, apprentice electricians, manufacturers' representatives, and electrical inspectors—just to name a few.

In fact, anyone involved in the electrical industry—in any capacity—will find helpful information in this book that will be of use on a daily basis, especially those electrical technicians involved with manufacturing processes.

The author is indeed grateful to all the manufacturers, power companies, and others who supplied reference material and many of the illustrations. Another bow of thanks goes out to Matt Fox of Prentice-Hall, Inc., who has worked with the author—through "thick and thin"—for over a decade on many other titles and somehow still manages to be a friend when needed.

John T. Earl

1

Introduction to Industrial Electrical Wiring

Industrial wiring installations include systems for all types of industrial plants, factories, refineries, and similar facilities. The basic principles of wiring for industrial installations are very similar to commercial installations except that in most cases the currents used in industrial electrical systems will be larger—requiring larger wire and conduit sizes; higher voltage will normally be used in industrial wiring systems; three-phase in addition to single-phase systems will be in use, and different types of materials and equipment may be involved.

FACTORS TO CONSIDER

Factors affecting the planning of an electrical installation for an industrial plant include the following:

1. New structure or modernization of an existing one.
2. Type of general building construction; for example, masonry, reinforced concrete, steel frame.
3. Type of floor, ceiling, partition, roof, and so on.
4. Type and voltage of service entrance, transformer connections, and whether service is underground or overhead.
5. Type and voltage of distribution system for power and lighting.

6. Type of required service equipment, such as unit substation, transformer bank.
7. Type of distribution system, including step-down transformers.
8. Who is responsible for furnishing the service-entrance and distribution equipment, the power company or plant?
9. Wiring methods, types of raceways, special raceways, busways, and the like.
10. Types of power-control equipment and extent of the worker's responsibility for connection to it.
11. Furnishing of motor starters, controls, and disconnects.
12. Extent of wiring to be installed on machine tools.
13. Extent of wiring connections to electric cranes and similar apparatus.
14. Type and construction of lighting fixtures, hangers and supports, and the like.
15. Extent of floodlighting. Type and dimensions of floodlighting supporting poles and mounting brackets.
16. Extent of signal and communication systems.
17. Ground conditions affecting the installation of underground wiring.
18. The size, type, and condition of existing wiring systems and services for modernization projects.
19. Whether the plant will be in use during the electrical installation.
20. Allowable working hours when an occupied building is being rewired.
21. Labor disputes.

Items 1 through 3 can be determined by studying the architectural drawings of the facility or, in the case of a modernization project, by a job site investigation.

Item 4 is usually worked out between the engineers and the local utility company, while items 5 through 10 are determined from the electrical drawings and specifications.

The electrical drawings as well as the mechanical drawings should be consulted to determine the condition of items 10 and 11. Also consult the special equipment section of the written specifications. In most cases, it is the responsibility of the trade furnishing motor-driven equipment to also furnish the starters and controls. The electrical contractor then provides an adequate circuit to each.

Item 12 can be determined by consulting the special equipment section of the written specifications and also by referring to the shop drawings supplied by the trade furnishing the equipment. It may be necessary to contact the manufacturer of the equipment in some cases.

On most projects, crane installation is a specialized category and installed by specialists in this field. All control work for the crane is normally done by the contractor who furnishes the crane. However, electrical workers will be required to furnish a feeder circuit from the main distribution panel to supply power for motors and controls to operate the crane. The extent of this work should be carefully coordinated between the trades involved.

Items 14 through 16 can be determined by consulting the electrical drawings and specifications for the project.

While items 17 through 19 might be covered to a certain extent in the project specifications, in most cases this information is obtained by a job site investigation.

Item 20 should be called out in the general specifications, but it may be necessary to hold a conference with the owners to determine exact conditions. Item 21 is determined by referring to the local labor agreement, by contacting the local union, or from past experience.

PLANNING AND COORDINATION

Even with carefully engineered drawings, the person in charge of the electrical installation in an industrial occupancy must still do much planning and coordination to carry out the work in the allotted amount of time. One problem that has existed in the past has been a variation of interpretation of the code requirements by two or more inspection authorities having coinciding jurisdiction over the same job. Therefore, at the planning stage, the superintendent or foreman should meet with all inspection authorities having jurisdiction to settle any problems at the outset.

Most industrial plants will have one or more wiring installations in hazardous locations. Therefore, those in charge should frequently consult the National Electrical (NE) Code to ascertain that all wiring is installed in a safe manner. Provisions must also be made to isolate the hazardous areas from those not considered hazardous.

Drawings from consulting engineers will vary in quality, and in most cases the wiring layout for a hazardous area is little different than the layout for a nonhazardous area. Usually the only distinction is a note on the drawing or in the specifications stating that the wiring in a given area or room shall conform to the NE Code requirements for hazardous locations. Rarely do the working drawings contain much detail of the system, leaving much of the design to the workers on the job. Therefore, the electrical foreman must study these areas very carefully, and consult the NE Code and other references, to determine exactly what is required.

Sometimes electrical contractors will have draftsmen prepare special drawings for use by their personnel in installing systems in such areas. If time permits, this is probably the best approach as it will save money plus much time in the field once the project has begun. In either case, whether preparing drawings or determining the requirements at the job site, considerable damage can be done to life and property if the system installed is faulty. Explosion-proof boxes, fittings, and equipment are very expensive and vary in cost for different types, sizes, and hub entrances, and require considerably more labor than nonhazardous installations.

For other than very simple systems, it is advisable to make detailed wiring layouts of all wiring systems in hazardous locations, even if it is only in the form of sketches and rough notes.

The typical plant wiring system will entail the connection of a power supply to many motors of different sizes and types. Sometimes the electrical contractor will be responsible for furnishing these motors, but in most cases they will be handled on special order from the motor manufacturers and will be purchased directly by the owners or by special equipment suppliers. Some factors involved in planning the wiring for electric motors include the following:

1. The type, size, and voltage of the motor and related equipment.
2. Who furnishes the motor, starter, control stations, and disconnecting means?
3. Is the motor separately mounted or an integral part of a piece of machinery or equipment?
4. Type and size of junction box or connection chamber on the motor.
5. The extent of control wiring required.
6. The type of wiring method of the wiring system to which the motor is to be connected, that is, conduit and wire, bus duct, trolley duct. Is the motor located in a hazardous area? If so, what provisions have been made to ensure that it will be wired according to the NE Code?
7. Who mounts the motor?
8. The physical shape and weight of the motor.

Obtaining all of the above information will facilitate the installation of all electric motors on the job by ensuring that proper materials will be available for the wiring and that no conflicts will arise between trades.

The installation of transformers and transformer vaults is another type of work that is frequently encountered in industrial wiring. A transformer vault, for example, is representative of the type of installation situation when, within a small area of the building and comprising a specialized section of the wiring system, a relatively small quantity of a number of different items of equipment and material is required. In many instances, even when working drawings and specifications are provided by consulting engineering firms, the vault will not be completely laid out to the extent that workers can perform the installation without further planning or questions. The major transformers, disconnects, and similar devices along with a one-line schematic diagram may be all that the drawings show. In such instances the foreman or workers must make a rough layout of the primary and secondary, indicating the necessary supports, supporting structures, connections, control and metering wiring, and the like. This calls for experienced knowledge and the ability to visualize the complete installation on the part of the person doing the layout work and supervision.

CABLE TRAY SYSTEMS

Cable tray systems are frequently used in industrial applications, and all electrical technicians involved in such work should be thoroughly familiar with the design and installation of such systems.

In general, a cable tray system must afford protection to life and property against faults caused by electrical disturbances, lightning, failures that are a part of the system, and failure of equipment that is connected to the system. For this reason, all metal enclosures of the system, as well as noncurrent-carrying or neutral conductors, should be bonded together and reduced to a common earth potential.

There is a frequent tendency to become lax in supplying the installation supervisor with definite layouts for a cable tray system. Under these conditions, considerable time is consumed in arriving at final decisions and definite routings before the work can proceed. On the other hand, it is often possible for one person to predetermine these layouts and save many hours of field erection time, provided careful planning is carried out.

For economical erection and satisfactory installation, working out the details of supports and hangers for the system is the job of the system designer and should not be left to the judgment of a field force not acquainted with the loads and forces to be encountered. Also, all types of supports and hangers should permit vertical adjustment, along with horizontal adjustment where possible. This can be accomplished by the use of channel framing, beam clamps, and threaded hanger rods.

All cable tray systems of appreciable size require a considerable quantity of hanger clips, support channels, hangers, and the like, in addition to cable tray items, which are intended for specific locations and may vary considerably in specifications. Too much emphasis, therefore, cannot be given to the necessity of packing, delivering, and receiving the material under definite and clear records so that the various items will be readily available when required, and also will not be confused with materials that are similar in appearance. All these factors should be considered during the planning stage.

When installing the cables proper precaution must be taken to avoid damaging the cables. A complete line of installation tools is available, developed through field experience, for pulling long lengths of cable up to 1000 ft or longer. These tools save considerable installation time.

Short lengths of cable can be laid in place without tools or can be pulled with a basket grip. Long lengths of small cable, 2 in. or less in diameter, can also be pulled with a basket grip. Larger cables, however, should be pulled by the conductor and the braid, sheath, or armor. This is done with a pulling eye applied at the cable factory or by taping the conductor to the eye of a basket grip and taping the tail end of the grip to the outside of the cable.

In general, the pull exerted on the cables pulled with a basket grip, not attached to the conductors, should not exceed 1000 lb. For heavier pulls, care should be taken not to stretch the insulation, jacket, or armor beyond the end of the conductor nor bend the ladder, trough, or channel out of shape.

The bending radius of the cable should not be less than the values recommended by the cable manufacturer, which range from 4 times the diameter for a rubber-insulated cable 1-inch maximum outside diameter without lead, shield, or armor, to 8 times the diameter for interlocked armor cable. Cables of special construction such as wire armor and high-voltage cables require a larger radius bend.

Best results are obtained in installing long lengths of cable up to 1000 ft with as many as a dozen bends by pulling the cable in one continuous operation at a speed of 20 to 25 ft/min. It may be necessary to brake the reel to reduce sagging of the cables between EZ rolls, which are devices designed especially for pulling cables in cable trays.

The pulling line diameter and length will, of course, depend on the pull to be made and construction equipment available. Winch and power unit should be of adequate size for the job and capable of developing the high pulling speed required for the best and most economical results.

Since cable tray installations are of major importance to industrial wiring systems, complete coverage of these systems are fully covered in another chapter in this book.

OTHER SYSTEMS

Other systems that are mainly used in industrial wiring applications and seldom found in residential or commercial installations include high-voltage substations, heavy-load generating plants, crane and hoists systems, ac and dc stand-by electrical systems, and enormous electric motor installations. These and other systems are thoroughly covered in later chapters.

For basic wiring methods, common to all electrical installations, the reader is referred to *Electrical Wiring: Design and Application*, by John Earl, and published by Prentice-Hall, Inc. Only methods and materials mainly used on industrial applications will be covered in this book.

2

Codes and Standards

Due to the potential fire and explosion hazards caused by the improper handling and installation of electrical systems, certain rules in the selection of materials, quality of workmanship, and precautions for safety must be followed. To standardize and simplify these rules and provide a reliable guide for electrical construction, the National Electrical Code (NE Code) was developed. This code, originally prepared in 1897, is frequently revised to meet changing conditions, improved equipment and materials, and new fire hazards. It is a result of the best efforts of electrical engineers, manufacturers of electrical equipment, insurance underwriters, fire fighters, and other concerned experts throughout the country.

The NE Code book is now published by the National Fire Protection Association, Batterymarch Park, Quincy, MA 02269. It contains specific rules and regulations intended to help safeguard persons and property from hazards arising from the use of electricity.

Although the NE Code itself states, "This Code is not intended as a design specification nor an instruction manual for the untrained person," it does provide a sound basis for the study of electrical design and installation procedures—under the proper guidance. The probable reason for the NE Code's self-analysis is that the code also states, "This Code contains provisions considered necessary for safety. Compliance therewith and proper maintenance will result in an installation essentially free from hazards, but not necessarily efficient, convenient, or adequate for good service or future expansion of electrical use."

The NE Code, however, has become the basic reference of the electrical construction industry, and anyone involved in electrical work, in any capacity, should obtain an up-to-date copy, keep it handy at all times, and refer to it frequently.

NATIONAL ELECTRICAL CODE CHAPTER OVERVIEW

In order that those involved in the electrical construction industry may understand the language and terms used in the NE Code, as well as in the industry in general, the definitions listed in Chapter 1, Article 100 of the code should be fully understood. For simplicity, this article lists only definitions essential to the proper use of the code; and then terms used in two or more other articles are defined in full. Other definitions, however, are defined in the individual articles of the NE Code where they apply.

General requirements for electrical installations are given in Article 110. This article, along with Article 100, should be read over several times until the information contained in both is fully understood and firmly implanted in the reader's mind. With a good understanding of this basic material, the remaining portions of the NE Code are easier to understand.

Chapter 2, Wiring Design and Protection, of the NE Code is the chapter that most electrical designers, workers, and others in the field will use the most. It covers such data as using and identifying grounded conductors, branch circuits, feeders, calculations, services, overcurrent protection, and grounding—all necessary for any type of electrical system, regardless of the building type in which the system is installed.

The chapter deals mainly with how-to items, for example, how to provide proper spacing for conductor supports, how to provide temporary wiring, how to size the proper grounding conductor or electrode. If a problem develops pertaining to the design or installation of a conventional electrical system, the answer can usually be found in this chapter.

Rules governing the wiring method and materials used in a specific installation are found in Chapter 3, Wiring Methods and Materials, of the NE Code. The provisions of this chapter apply to all wiring installations except remote-control switching, low-energy power, signal systems, communication systems, and conductors that form an integral part of equipment, such as motors and motor controllers.

Rules pertaining to raceways, boxes, cabinets, and raceway fittings are also found in Chapter 3 of the NE Code. Since outlet boxes vary in shape to accommodate the size of the raceway, the number of conductors entering the box, the type of building construction, the atmospheric condition of the building, and other special requirements, this chapter is designed to answer most questions that might arise.

Chapter 3 of the NE Code also generally covers a wide variety of switches, push buttons, pilot lamp receptacles, and convenience outlets, as well as switchboards and panelboards. Such items as location, installation methods, clearances, grounding, and overcurrent protection are thoroughly covered in this section of the code.

Chapter 4, Equipment for General Use, of the NE Code begins with the use and installation of flexible cords and cables, including the trade name, type letter, wire size, number of conductors, conductor insulation, outer covering, and use of each. The articles continue on to fixture wires, again giving trade name, type letter, and other pertinent details.

The article on lighting fixtures in Chapter 4 is of special interest to electricians and designers as it gives the installation procedures for the various types of fixtures for use in specific locations (fixtures near combustible materials, in closets, etc.).

The selection of electric motors is found in Articles 430 through 445 of the NE Code, as well as mounting the motor and making electrical connections to it. Heating equipment, transformers, and capacitors are also found in Chapter 4 of the NE Code.

While storage batteries are not often thought of as part of an electrical wiring system for building construction, they are often used to provide standby emergency lighting service and to supply power to security systems separate from the main ac electrical system. Most requirements pertaining to battery-operated systems will be found in Chapter 4 of the NE Code.

Any location where the atmosphere or material in the area is such that the sparking of operating electrical equipment may cause an explosion or fire is considered a hazardous location, and these areas are covered in Chapter 5 of the NE Code, Special Occupancies. These locations have been classified in the NE Code into certain class locations, and various atmospheric groups have been established on the basis of the explosive character of the atmosphere for the testing and approval of equipment for use in these various groups.

The basic principle of explosion-proof wiring is to design and install a system so that, when the inevitable arcing occurs within the electrical system, ignition of the surrounding explosive atmosphere is prevented. The basic principles of such an installation are covered in Chapter 5 of the NE Code.

Theaters and similar occupancies also fall under the regulations set forth in Chapter 5 of the NE Code. Recognizing that hazards to life and property due to fire and panic exist in theaters, there are certain requirements in addition to those of the usual commercial wiring installations. While drive-in-type theaters do not present the inherent hazards of enclosed auditoriums, the projection rooms and other areas adjacent to these rooms must be properly ventilated and wired for the protection of operating personnel and others using the area.

Other areas falling under the regulations of Chapter 5 of the NE Code include residential storage garages, aircraft hangars, service stations, bulk-storage plants, finishing processes, health-care facilities, mobile homes and parks, and recreation vehicles and parks.

The provisions in Chapter 6, Special Equipment, of the NE Code apply to electric signs and outline lighting, cranes and hoists, elevators, electric welders, and sound-recording and similar equipment. Therefore, any electrical designer or electrician who works on any of these systems should thoroughly check through Chapter 6 of the NE Code, along with other chapters that may apply.

Electrical signs and outline lighting are usually considered to be self-contained equipment installed outside and apart from the building wiring system. However, the circuits feeding these lights are usually supplied from within the building itself. Therefore, some means of disconnecting such equipment from the supply circuit is required. All such equipment must be grounded except when insulated from the ground or conducting surfaces or if inaccessible to unauthorized persons.

Neon tubing, where used, requires the use of step-up transformers to provide the necessary operating voltages, and secondary conductors must have insulation and be rated for this high voltage; terminators must also be of the proper type and must be protected from or inaccessible to unqualified persons. In most instances, the electrician will only be responsible for providing the feeder circuit to the location of the lights; qualified sign installers will usually do the actual sign work.

Cranes and hoists are usually furnished and installed by those other than electricians. However, it is usually the electrician's responsibility to furnish all wiring, feeders, and connections for the equipment. Such wiring will consist of the control and operating circuit wiring on the equipment itself and the conductors supplying electric current to the equipment in a manner to allow it to move or operate properly. Furthermore, electricians are normally required to furnish and install the contact conductor and necessary suspension and supporting insulators. Motors, motor-control equipment, and similar items are normally furnished by the crane manufacturer.

The majority of the electrical work involved in the installation and operation of elevators, dumbwaiters, escalators, and moving walks is usually furnished and installed by the manufacturer. The electrician is usually only required to furnish a feeder terminating in a disconnect means in the bottom of the elevator shaft and perhaps to provide a lighting circuit to a junction box midway in the elevator shaft for connection of the elevator-cage lighting cable. Articles in Chapter 6 of the NE Code will give most of the requirements for these installations.

Electric welding equipment is normally treated as a piece of industrial power equipment for which a special power outlet is provided. Certain specific conditions, however, apply to circuits supplying welding equipment and are outlined in Chapter 6 of the NE Code.

Wiring for sound-recording and similar equipment is essentially of the low-voltage type. Special outlet boxes or cabinets are usually provided with the equipment, although some items may be mounted in or on standard outlet boxes. Some systems of this type require direct current, which is obtained from rectifying equipment, batteries, or motor generators. The low-voltage alternating current is obtained through the use of relatively small transformers connected on the primary side to a 120-V circuit within the building.

Other items covered under Chapter 6 of the NE Code include x-ray equipment, induction and dielectric heat-generating equipment, and machine tools. A brief reading of this chapter will provide a sound basis for approaching such work in a professional manner, and then problems can be studied in more detail during installation.

In most commercial buildings, codes and ordinances require a means of lighting public rooms, halls, stairways, and entrances so that, if the general building lighting is interrupted, there will be sufficient light to allow the occupants to exit from the building. Exit doors must also be clearly indicated by illuminated exit signs.

Articles in Chapter 7, Special Conditions, of the NE Code give provisions for installing emergency lighting systems. Such circuits should be arranged so that they may be automatically transferred over to an alternate source of current supply (from storage batteries, gasoline-driven generators, or properly connected to the supply side of the main service) so that disconnecting the main service switch will not disconnect the emergency circuits. Additional details may be found in Article 700 of the NE Code.

Circuits and equipment operating at more than 600 V between conductors will also be found in this portion of the code. In general, conductors' insulations must be of a type approved for the operating voltage, and the conductors must be installed in rigid conduit, duct, or armored cable approved for the voltage used. These cables must also be terminated with approved cable-terminating devices. All exposed live parts must be given careful attention and adequately guarded by suitable enclosures or isolated by elevating the equipment beyond the reach of unauthorized personnel. Overcurrent protection and disconnecting means must be manufactured specifically for the operating voltage of the system. Examples of such high-voltage installations include feeders for synchronous motors and condensers, substations and transformer volts, and the like. Other details are covered in articles in Chapter 7 of the NE Code.

Among other items in Chapter 7 of the NE Code is the installation of outside wiring other than for electric signs. Such wiring is either attached to the building or between two or more buildings, run overhead, underground, or in a raceway fastened to the face of the building.

Overhead systems may consist of individual conductors supported on or by insulators on or at the building surface, or by means of approved cable either attached to or suspended between buildings. When the buildings are some distance from each other, intermediate supporting poles are necessary, and certain clearance distances must be maintained over driveways and buildings.

Underground wiring between buildings or outside buildings is installed either directly in the ground as direct-burial cable or else pulled through raceways consisting of rigid conduit, PVC conduit, or ducts encased in concrete.

Chapter 8, Communication Systems, of the NE Code deals with communication systems and circuits, that is, telephone, telegraph, district messenger, fire and burglar alarms, and similar central station systems and telephone systems not connected to a central station system but using similar types of equipment, methods of installation, and maintenance. Articles in this chapter also cover radio and television equipment, community antenna television and radio distribution systems (cable TV), and similar systems. The basic requirements are outlined in this chapter of the code, but a good knowledge of communications systems is also required for a proper installation.

Local inspection authorities frequently review actual construction sites to make a determination to grant exceptions for the installation of conductors and equipment not under the exclusive control of the utility or power companies and used to connect the electric utility supply system to the service-entrance conductors of the premises served, provided such installations are outside a building or terminate immediately inside a building wall.

Furthermore, the authority having jurisdiction on a particular project may waive specific requirements in the NE Code or permit alternate methods where it is assured that equivalent objectives can be achieved by establishing and maintaining effective safety procedures.

The NE Code is also intended to be suitable for mandatory application by government bodies exercising legal jurisdiction over electrical installations and for use by insurance inspectors. The authority having jurisdiction for enforcement of the NE Code will have the responsibility for making interpretations of the rules, for deciding on the approval of equipment and materials, and for granting the special permission contemplated in a number of the rules.

LOCAL CODES AND ORDINANCES

A number of towns, cities, and counties have their own local electrical code or ordinance. In general, these are based on, or similar to, the NE Code, but on certain classes of work, they may have a few specific rules that are usually more rigid than the NE Code.

In addition to the NE Code and local ordinances of certain cities, local power companies may have some special rules regarding location of service-entrance wires, watt-hour meter connections, and similar details that must be satisfied before connection can be made to a building.

Mandatory rules of the NE Code are characterized by the use of the word *shall*, while advisory rules are characterized by the use of the word *should*. When statements are made using the latter word, they are stated as recommendations which are advised but not necessarily required. Some local ordinances, however, may change should to shall. When working in a new area, it is therefore useful to find out if there are legal requirements amending the NE Code and, specifically, what these are.

REQUIREMENTS OF ARCHITECTS AND ENGINEERS

When any type of building of any consequence is contemplated, an architect is usually commissioned to prepare the complete working drawings and written specifications for the building. For projects of any consequence, the architect usually includes drawings and specifications for the complete electrical system, which are usually prepared by consulting engineering firms.

Consulting engineers will often specify materials and methods that surpass the requirements of the NE Code in order to obtain a high-quality finished job. Electricians who work on such jobs must comply with the working drawings to carry out the engineer's design. For example, number 14 American wire gauge (AWG) wire may be quite adequate for a certain wiring installation according to the NE Code. However, the engineer may specify that wire smaller than number 12 AWG cannot be used on a particular project. If so, the installation must be carried out as specified, even though it surpasses the NE Code. For such reasons, all persons involved in a building construction project should carefully study the working drawings and construction documents before starting the wiring installation and refer to them often as the work progresses.

Reading electrical drawings is covered in Chapter 3 of this book. You may also want to obtain a copy of *Electrical Blueprint Reading*, published by Craftsman Book Company, 6058 Corte Del Cedro, Carlsbad, CA 92008.

TESTING LABORATORIES

There are several qualified testing laboratories in the United States and Canada, but Underwriters Laboratories (UL), Inc., is the most widely used. It investigates, studies, experiments, and tests products, materials, and systems. If the items are found to meet the UL safety requirements during these tests, UL will list the items. Bear in mind, however, that UL does not approve such items; it only lists them as having passed their tests for safety. Therefore, the phrase *listed by UL* is the correct one rather than approved by UL.

Tested products that meet UL standards are listed under various categories in directories published by UL. They can be purchased from Underwriters Laboratories, Inc., 333 Pfingsten Road, Northbrook, IL 60062.

Either the products or the containers in which the products are shipped, or both, should contain the UL listing mark. The listing mark may be on an interior or exterior surface of the product, and some will have the mark on the shipping carton or reel.

The electrician should become familiar with this mark and keep in mind that many electrical inspectors and consulting engineers will allow only UL-listed items to be installed in an electrical system. The failure to do so could cost someone considerable time and expense in replacing the nonlisted items.

3

Blueprint Reading

Those involved with industrial electrical systems in any capacity will encounter many types of drawings and diagrams. For example, the engineer or designer will need to study the layout of areas to be lighted to obtain important information for laying out lighting fixtures and to calculate illumination requirements. Furthermore, the engineer or designer must be able to make sketches so the draftsmen may complete working drawings for the workers who perform the installation and maintenance of the system. Draftsmen must be able to read blueprints so they can interpret the engineer's sketches, and workers on the job must be able to read drawings and diagrams so that the various jobs are correctly performed. Therefore, a brief sampling of the various types of drawings that may be encountered in the electrical field is in order.

PICTORIAL DRAWINGS

In this type of drawing the objects are drawn in one view only; that is, three-dimensional effects are simulated on the flat plane of drawing paper by drawing several faces of an object in a single view. This type of drawing is very useful to describe objects and convey information to those who are not well trained in blueprint reading or to supplement conventional diagrams in certain special cases.

One example of a pictorial drawing would be an exploded view of a motor starter used to control a fan-coil unit to show the physical re-

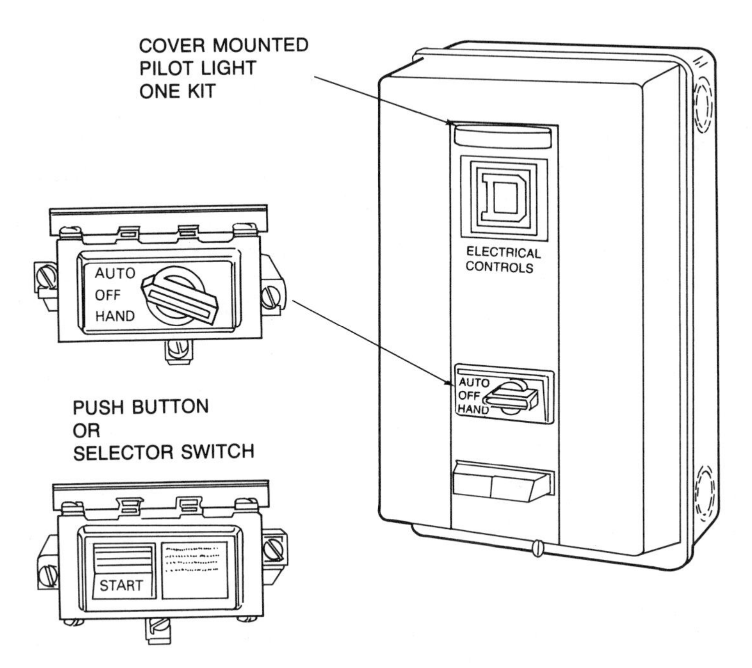

Figure 3-1 Pictorial drawing of a motor starter.

lationship of each part so that the starter could be disassembled and reassembled during maintenance. See Fig. 3-1.

The types of pictorial drawings most often found in the heating, ventilating, and cooling (HVAC) industry include:

1. Isometric
2. Oblique
3. Perspective

All of these drawings are relatively difficult to draw and their use is normally limited to the manufacturers of heating components for showing their products in catalogs, brochures, and similar publications. However, in recent times, they are gradually being replaced by photographs where possible.

By definition, an isometric drawing is a view projected onto a vertical plane in which all of the edges are foreshortened equally. Figure 3-2 shows an isometric drawing of a cube. In this view, the edges are 120° apart and are called the isometric axes, while the three surfaces shown are called the isometric planes. The lines parallel to the isometric axes are called the isometric lines.

Isometric drawings are usually preferred over the other two types mentioned for use in engineering departments to show certain details

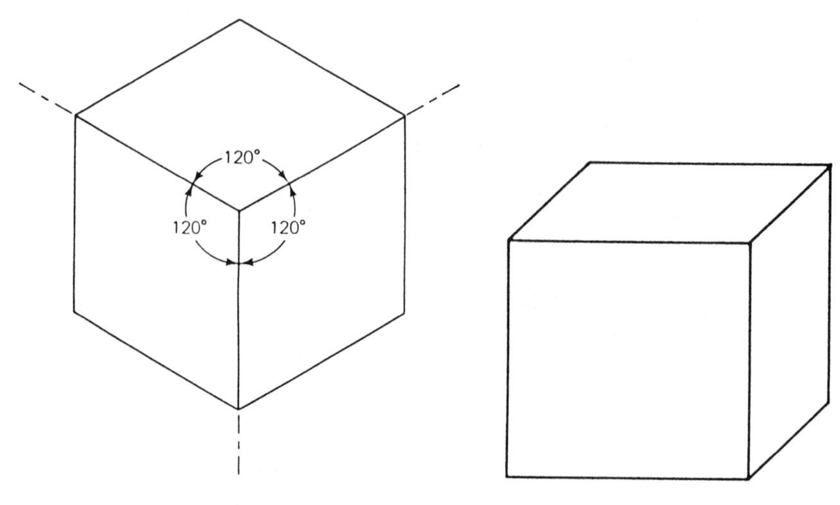

Figure 3-2 Isometric drawing of a cube.

Figure 3-3 Oblique drawing of a cube.

on installation drawings, because it is possible to draw isometric lines to scale with a 30-60° triangle.

The oblique drawing is similar to the isometric drawing in that one face of the object is drawn in its true shape and the other visible faces are shown by parallel lines drawn at the same angle (usually 45-30°) with the horizontal. However, unlike an isometric drawing, the lines drawn at a 30° angle are shortened to preserve the appearance of the object and are therefore not drawn to scale. The drawing in Fig. 3-3 is an oblique drawing of a cube.

The two methods of pictorial drawing described so far produce only approximate representations of objects as they appear to the eye, as each type produces some degree of distortion of any object so drawn. However, because of certain advantages, the previous two types are the ones most often found in engineering drawings.

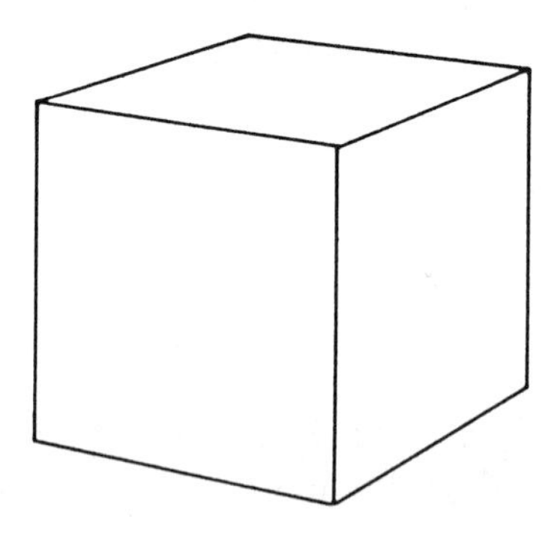

Figure 3-4 Perspective drawing of a cube.

Sometimes—for a certain catalog illustration, or a more detailed instruction manual—one wants to draw an exact pictorial representation of a object as it actually appears to the eye. A drawing of this type is called a perspective drawing; one such drawing—again of the cube—appears in Fig. 3-4.

ORTHOGRAPHIC-PROJECTION DRAWINGS

An orthographic-projection drawing is one that represents the physical arrangement and views of specific objects. These drawings give all plan views, elevation views, dimensions, and other details necessary to construct the project or object. For example, Fig. 3-5 suggests the form of a block, but it does not show the actual shape of the surfaces, nor does it show the dimensions of the object so that it may be constructed.

An orthographic projection of the block in Fig. 3-5 is shown in Fig. 3-6. One of the drawings in this figure shows the block as though the observer were looking straight at the front; one, as though the observer were looking straight at the left side; one, as though the observer were looking straight at the right side; and one, as though the observer

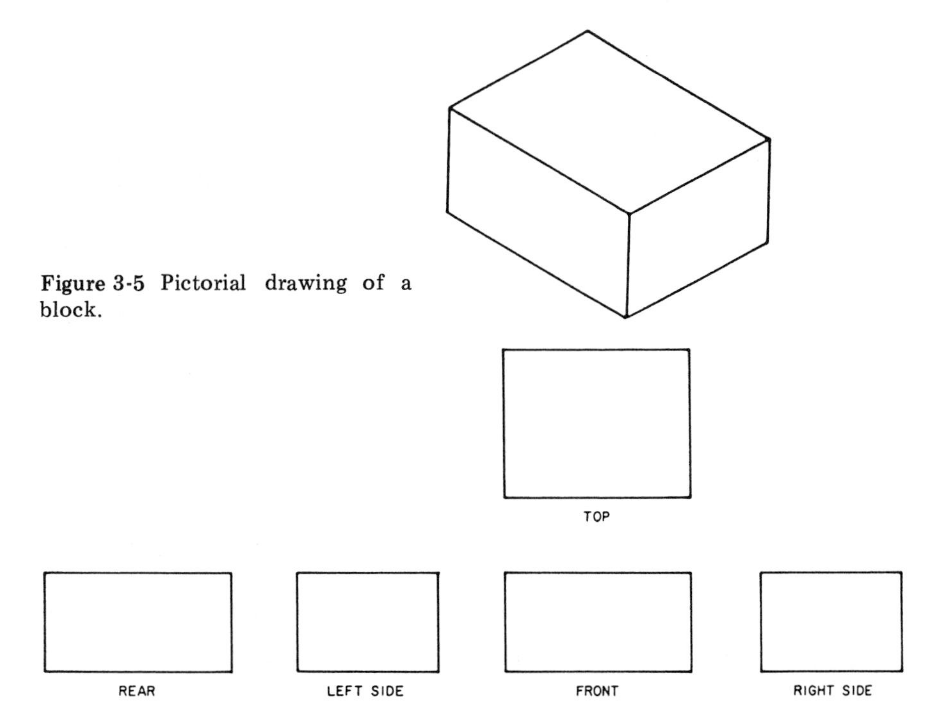

Figure 3-5 Pictorial drawing of a block.

Figure 3-6 Orthographic projection of the block in Fig. 3-5.

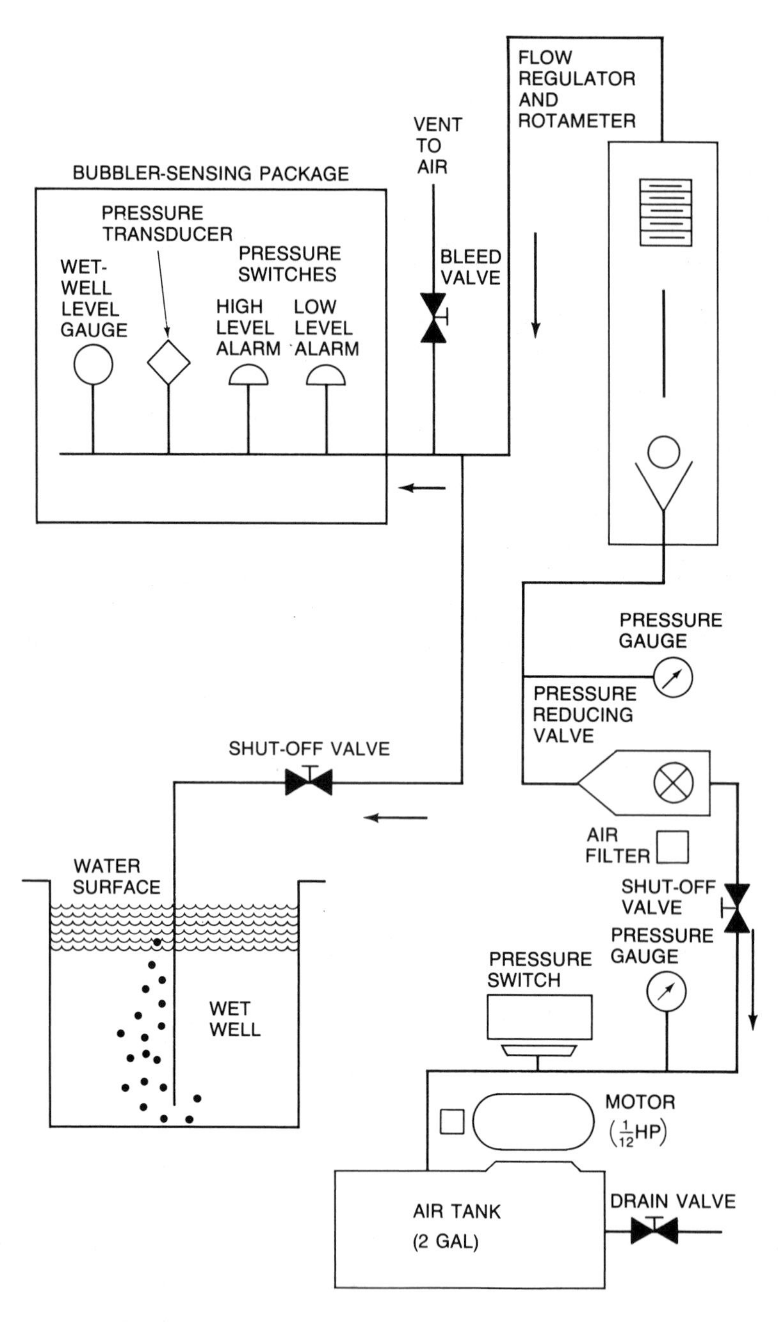

Figure 3-7 Typical flow diagram showing a bubbler.

were looking at the rear of the block. The remaining view is as if the observer were looking straight down on top of the block. These views, when combined with dimensions, will allow the object to be constructed properly from materials such as metal, wood, plastic, or whatever the specifications call for.

ELECTRICAL DIAGRAMS

Electrical diagrams are drawings intended to show, in diagrammatic form, electrical components and their related connections. Such drawings are seldom drawn to scale, and show only the electrical association of the different components. In diagram drawings, symbols are used extensively to represent the various pieces of electrical equipment or components, and lines are used to connect these symbols—indicating the size, type, number of wires, and the like.

In general, the types of diagrams that will be encountered by those working with heating systems will include flow diagrams (Fig. 3-7), single-line block diagrams (Fig. 3-8), and schematic wiring diagrams (Fig. 3-9). All of these types are frequently found in HVAC control diagrams.

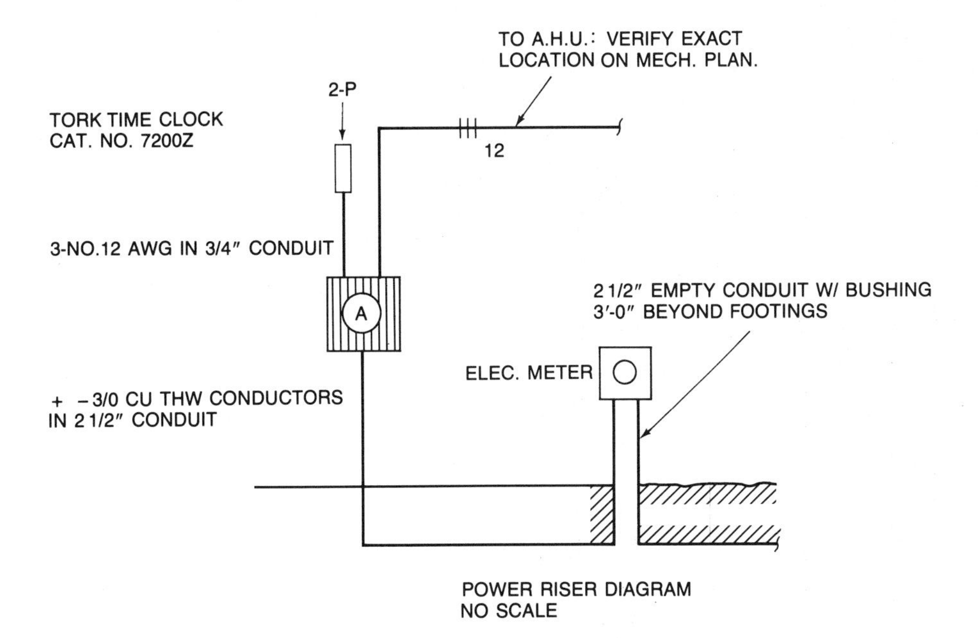

Figure 3-8 Single-line block diagram.

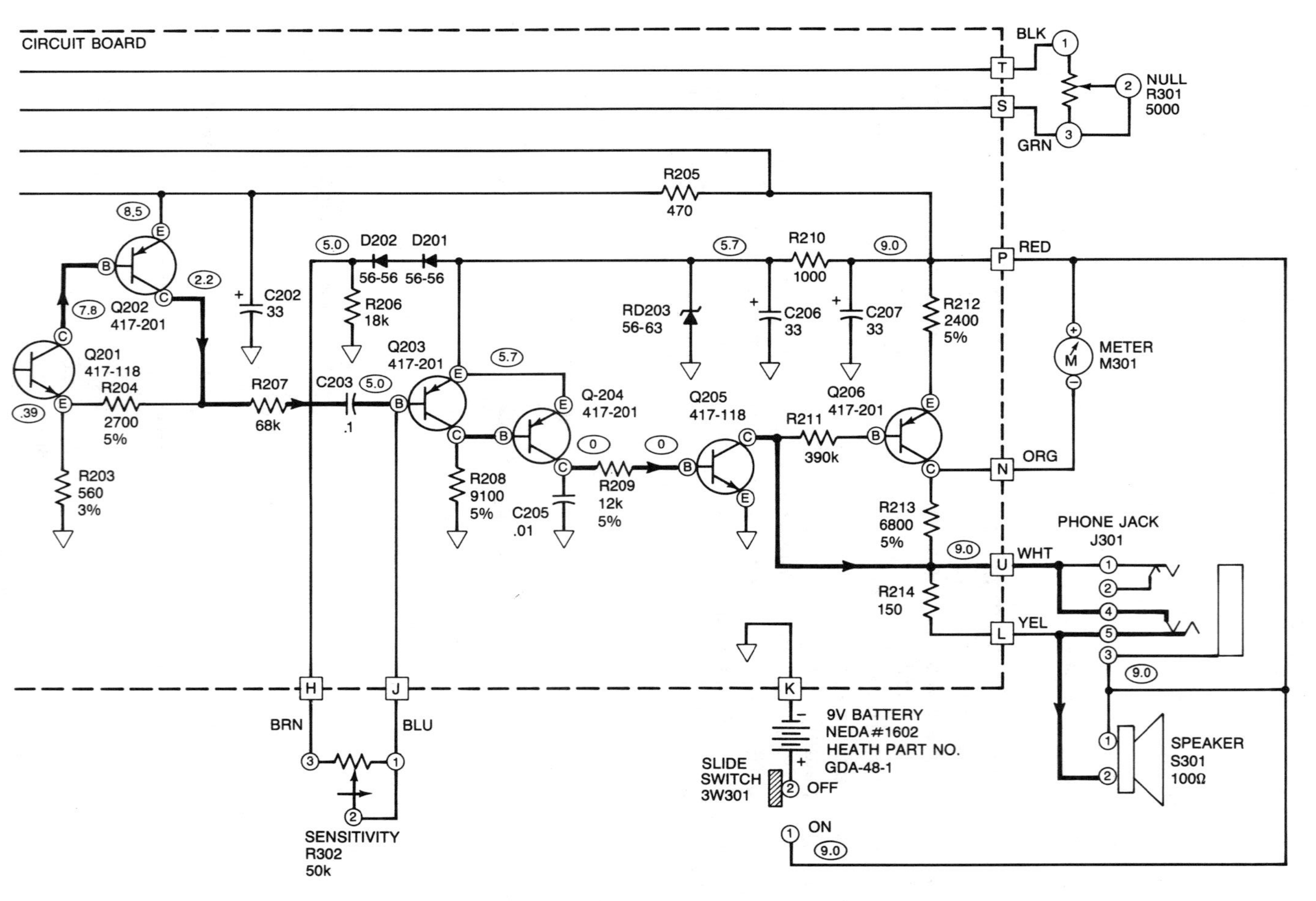

Figure 3-9 Schematic wiring diagram.

ELECTRICAL/ELECTRONIC GRAPHIC SYMBOLS

The purpose of a "working" drawing—as applied to the electrical heating industry—is to show how a certain object, piece of equipment, or system is to be constructed, installed, modified, or repaired. An electronic testing instrument, for example, usually has drawings and specifications showing the mechanical arrangement of the chassis and housing, and a schematic diagram showing the various components, the power supply, and the connections between each. An electrical drawing of a system used in a building to indicate the routing of the control circuits usually shows the floor plans of each level, the routing of the conduit or conductors, the number and sizes of wires or cables, heating equipment, feeders, and other information for the proper installation of the system.

In preparing drawings for the electrical industry, symbols are used to simplify the work of those preparing the drawing. To illustrate this fact, look at the pictorial drawing of a motor starter in Fig. 3-10. Although this drawing clearly indicates the type of control, method of connections, and the like, such drawings would take hours for a draftsman to complete, costing more than could possibly be allotted for the conventional working drawing used in electrical construction. However, by using a drawing such as the one in Fig. 3-11, which utilizes symbols to indicate the various components, the drafting time can be cut back to only minutes, and to the experienced worker, both drawings relay the exact same information.

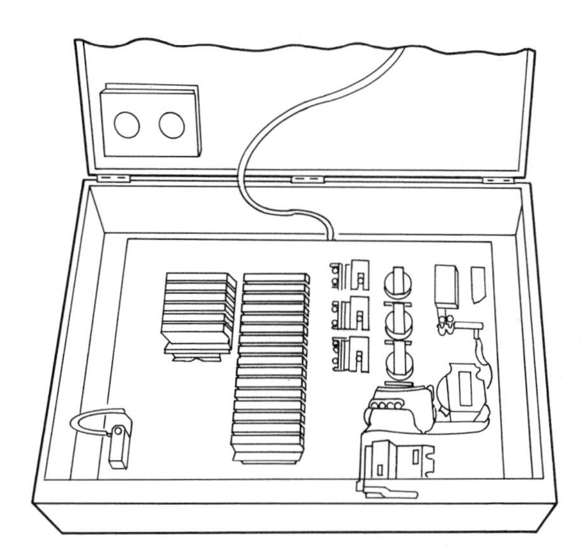

Figure 3-10 Pictorial drawing of a motor starter.

Figure 3-11 Motor starter represented via drawing symbols.

Most engineers, designers, and draftsmen use symbols adopted by the United States of America Standards Institute (USASI) for use on electrical and electronic drawings [Fig. 3-12(a)]. However, many designers and draftsmen frequently modify these symbols to suit their own particular requirements for the type of work they normally encounter. For this reason, most working drawings will have a symbol list or legend placed on the drawing to describe exactly what each symbol means—eliminating practically any doubt as to what is exactly required. A typical symbol list or legend appears in Fig. 3-12(b).

It is evident from the list in Fig. 3-12 that many symbols have the same basic form, but their meanings differ slightly because of the addition of a line, mark, or abbreviation. Therefore, a good procedure to follow in learning the different electrical symbols is to first learn the basic form and then apply the variations of that form to obtain the different meanings.

Note also that some of the symbols in Fig. 3-12 are abbreviations, such as XFER for transfer and WT for watertight. Others are simplified pictographs, such as for externally operated disconnect switch, or for a nonfusible safety switch, using both pictographs and abbreviations.

The most common abbreviations found in control diagrams include the following:

1. SP — Single Pole
2. ST — Single Throw
3. DP — Double Pole
4. DT — Double Throw
5. 3P — Three Pole
6. 2P — Two Pole
7. NC — Normally closed contact
8. NO — Normally open contact

Normally closed contact means the contact is closed when the relay coil is not energized. Normally open contact means the contact is open when the relay coil is not energized. Contacts will change position when the relay coil is energized. The normally closed contacts will open and the normally open contacts will close.

Electrical Reference Symbols

ELECTRICAL ABBREVIATIONS
(Apply only when adjacent to an electrical symbol.)

Central Switch Panel	CSP
Dimmer Control Panel	DCP
Dust Tight	DT
Emergency Switch Panel	ESP
Empty	MT
Explosion Proof	EP
Grounded	G
Night Light	NL
Pull Chain	PC
Rain Tight	RT
Recessed	R
Transfer	XFER
Transformer	XFRMR
Vapor Tight	VT
Water Tight	WT
Weather Proof	WP

ELECTRICAL SYMBOLS

Switch Outlets

Single-Pole Switch	S
Double-Pole Switch	S_2
Three-Way Switch	S_3
Four-Way Switch	S_4
Key-Operated Switch	S_K
Switch and Fusestat Holder	S_{FH}
Switch and Pilot Lamp	S_P
Fan Switch	S_F
Switch for Low-Voltage Switching System	S_L
Master Switch for Low-Voltage Switching System	S_{LM}
Switch and Single Receptacle	⊖S
Switch and Duplex Receptacle	⊜S
Door Switch	S_D
Time Switch	S_T
Momentary Contact Switch	S_{MC}
Ceiling Pull Switch	Ⓢ
"Hand-Off-Auto" Control Switch	[HOA]
Multi-Speed Control Switch	[M]
Push Button	[•]

Receptacle Outlets

Where weather proof, explosion proof, or other specific types of devices are to be required, use the upper-case subscript letters. For example, weather proof single or duplex receptacles would have the uppercase WP subscript letters noted alongside of the symbol. All outlets should be grounded.

Single Receptacle Outlet	⊖
Duplex Receptacle Outlet	⊜
Triplex Receptacle Outlet	
Quadruplex Receptacle Outlet	

Figure 3-12 (a) Electrical symbols used on working drawings, (b) electrical symbol list.

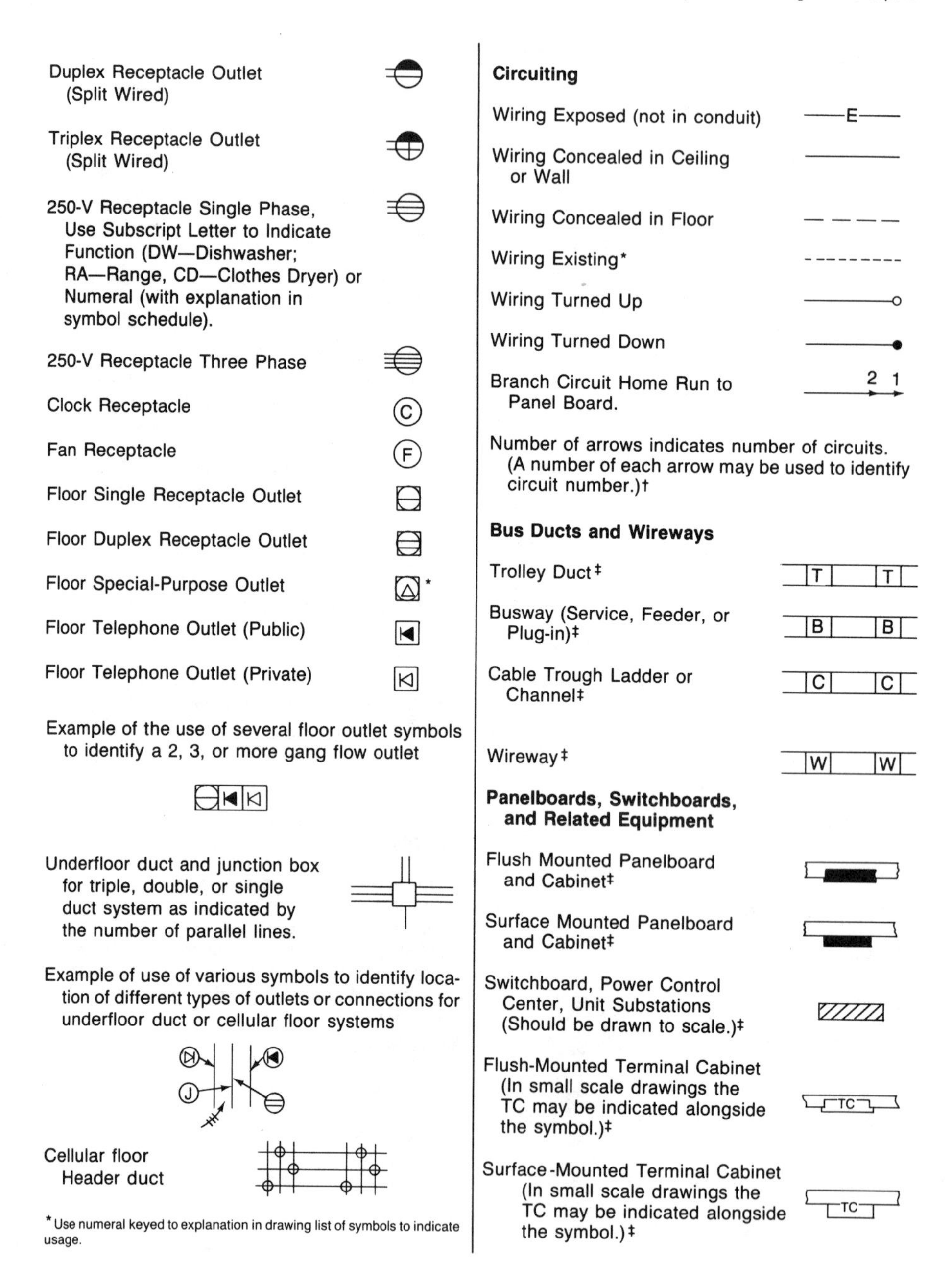

Figure 3-12 (cont.)

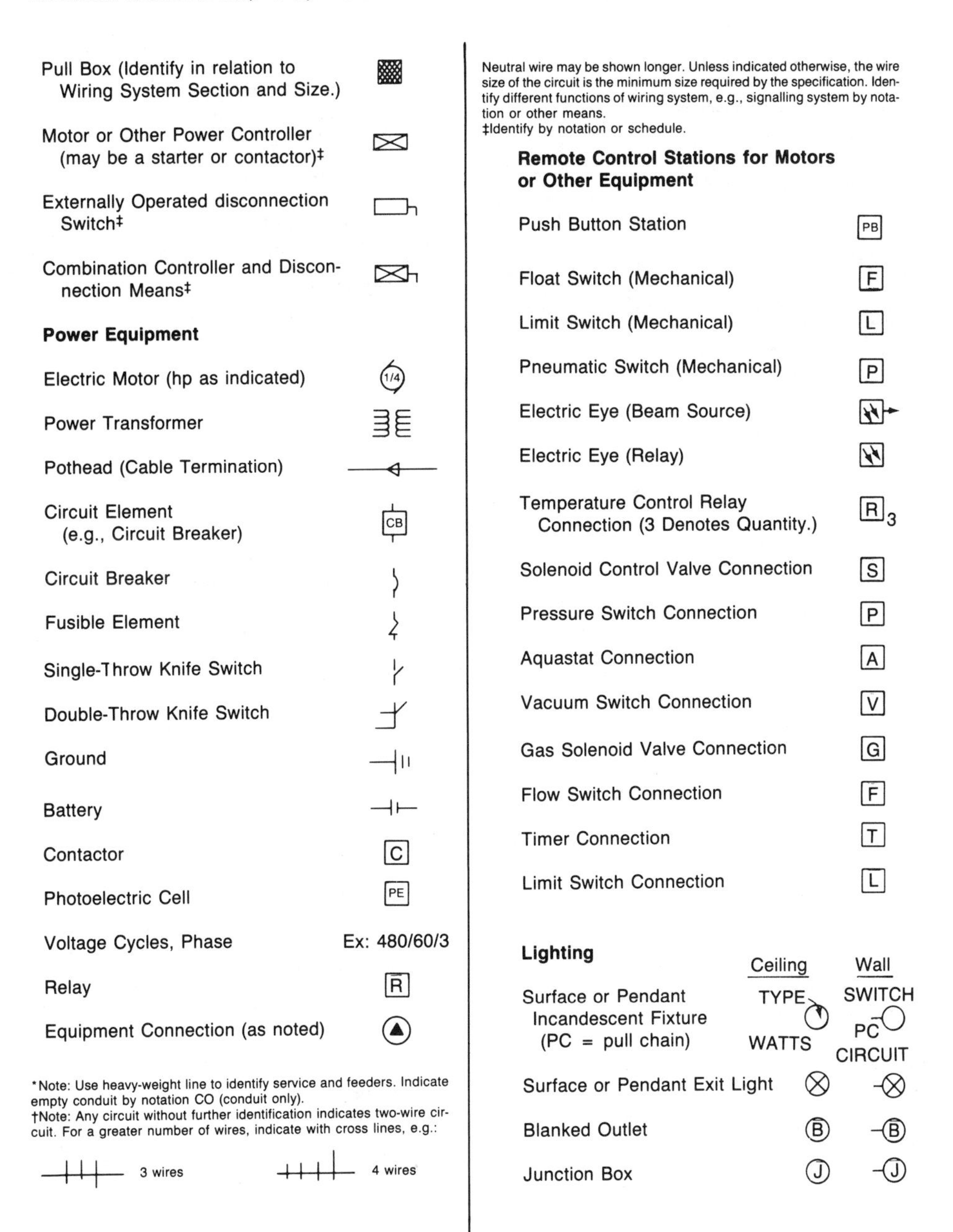

Figure 3-12 (cont.)

Recessed Incandescent Fixtures

Surface or Pendant Individual Fluorescent Fixture

Surface or Pendant Continuous-Row Fluorescent Fixture (Letter indicating controlling switch)

A
1/100 ← Fixture No. ← Wattage

Bare-Lamp Fluorescent Strip*

*In the case of continuous-row bare-lamp flourescent strip above an area-wide diffusing means, show each fixture run using the standard symbol; indicate area of diffusing means and type by light shading and/or by light shading and/or drawing notation.

Electric Distribution or Lighting System, Aerial

Pole‡

Steel or Parking Lot Light and Bracket‡

Transformer‡

Primary Circuit‡

Secondary Circuit‡

Down Guy

Head Guy

Sidewalk Guy

Service Weather Head‡

Electric Distribution or Lighting System, Underground

Manhole‡ M

Handhole‡ H

Transformer Manhole or Vault‡ TM

Transformer Pad‡ TP

Underground Direct Burial Cable (Indicate type, size, and number of conductors by notation or schedule)

Underground Duct Line (Indicate type, size, and number of ducts by cross-section identification of each run by notation or schedule. Indicate type, size, and number of conductors by notation or schedule.

Street Light Standard Feed From Underground Circuit‡

‡Identify by notation or schedule.

Signalling System Outlets

Institutional, Commercial, and Industrial Occupancies

I. Nurse Call System Devices (any type)

Basic Symbol

(Examples of individual item identification. Not a part of standard.)

Nurses' Annunciator (Adding a number after it indicates number of lamps e.g., ① 24)

Call Station, Single Cord, Pilot Light ②

Call Station, Double Cord, Microphone Speaker ③

Corridor Dome Light, 1 Lamp ④

Transformer ⑤

Any other item on same system (use numbers as required). ⑥

II. Paging System Devices (any type)

Basic Symbol

Figure 3-12 (cont.)

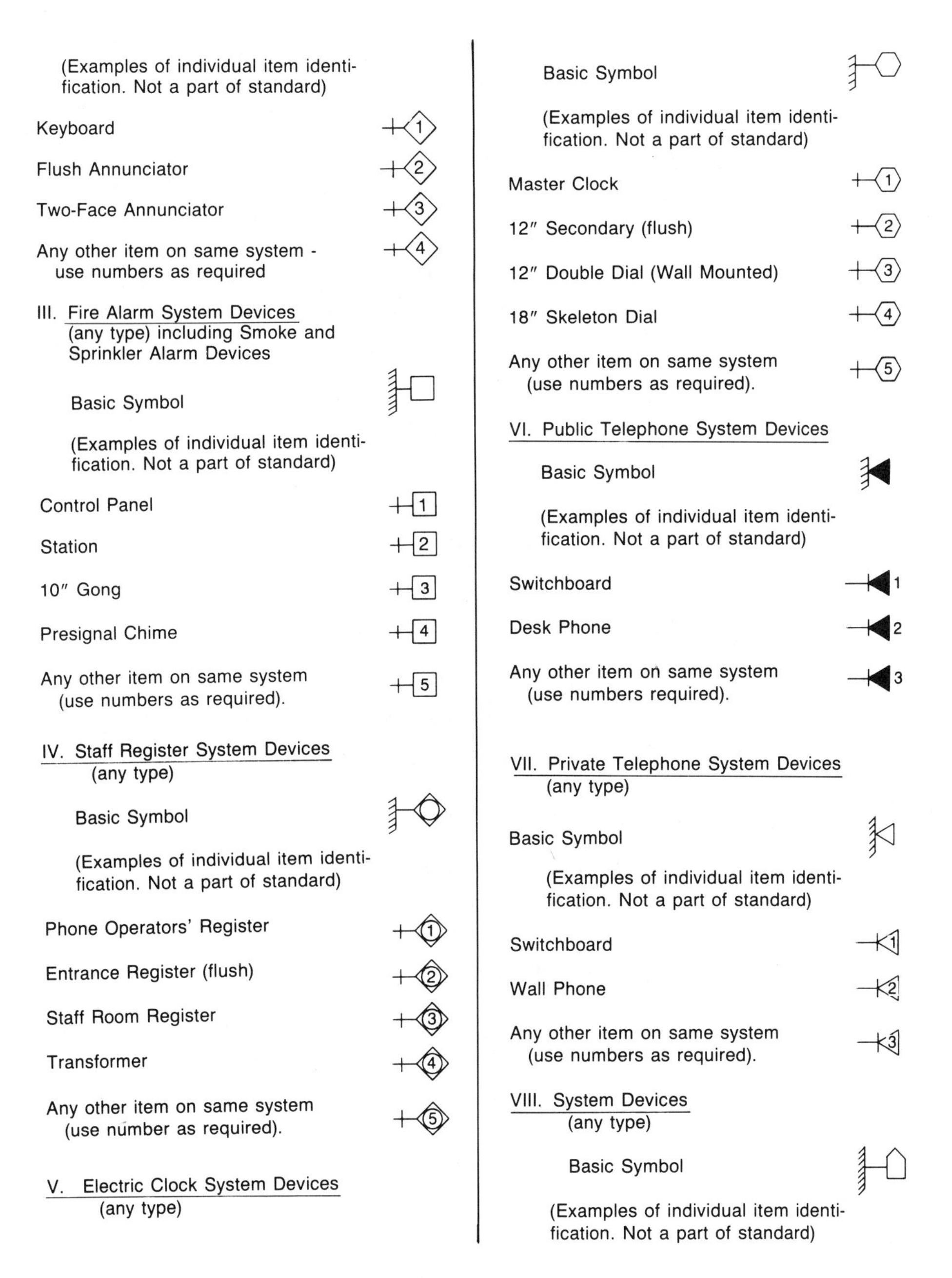

Figure 3-12 (cont.)

Central Station

Key Station

Any other item on same system
(use numbers as required).

IX. Sound System

Basic Symbol

(Examples of individual item identification. Not part of standard)

Amplifier

Microphone

Interior Speaker

Exterior Speaker

Any other item on same system
(use numbers as required).

X. Other Signal System Devices

Basic Symbol

(Examples of individual item identification. Not a part of standard)

Buzzer

Bell

Pushbutton

Annunciator

Any other item on same system
(use numbers as required).

Residential Occupancies

Signalling system symbols for use in identifying standardized residential-type signal system items on residential drawings where a descriptive symbol list is not included on the drawing. When other signal system items are to be identified, use the above basic symbols for such items together with a descriptive symbol list.

Push Button

Buzzer

Bell

Combination Bell–Buzzer

Chime — CH

Annunciator

Electric Door Opener — D

Maid's Signal Plug — M

Interconnection Box

Bell-Ringing Transformer — BT

Outside Telephone

Interconnecting Telephone

Television Outlet — TV

Figure 3-12 (cont.)

Electrical Symbol List

NOTE: These Are Standard Symbols and May Not All Appear on the Project Drawings; However, Wherever the Symbol on Project Drawings Occurs, the Item Shall be Provided and Installed.

Symbol	Description
	Ceiling Outlet with Incandescent Fixture
	Recessed Outlet with Incandescent Fixture
	Wall-Mounted Outlet with Incandescent Fixture
	Ceiling Outlet with Fluorescent Fixture
	Wall-Mounted Outlet with Fluorescent Fixture
	Fluorescent Fixture Mounted Under Cabinet
	Ground-Mounted Uplight
	Post-Mounted Incandescent Fixture
	Floodlight Fixture
	Fluorescent Strip
	Exit Light, Surface or Pendant
	Exit Light, Wall Mounted
5	Indicates Type of Lighting Fixture (see schedule)
S	Single-Pole Switch Mounted 50″ up to ℄ of Box
S_3	Three-Way Switch Mounted 50″ up to ℄ of Box
S_4	Four-Way Switch Mounted 50″ up to ℄ of Box
S_2	Two-Pole Switch Mounted 50″ up to ℄ of Box
S_L	Low-Voltage Switch to Relay
S_D	Door Switch
	Duplex Receptacle Mounted 18″ Up to Center of Box
	Duplex Receptacle Mounted 4″ Above Countertop
	Split-Wired Duplex Receptacle (top half switched)
L	Special Outlet or Connection—Numeral Indicates Type (see legend at end of list)
H	Floor-Mounted Receptable
C	Clock-Hanger Receptacle
	Pushbutton Switch for Door Chimes
CH	Chimes

Figure 3-12 (cont.)

Symbol	Description
TV	TV Outlet Mounted 18″ up to ℄ of Box
	Telephone Outlet
	Fusible Safety Switch
N	Nonfusible Safety Switch
	Main Distribution Panel
	Lighting (Panel Numeral Indicates Type)
	Branch Circuit Concealed in Ceiling or Walls (Slash Marks Indicate Number of Conductors in Run. Two Conductors Not Noted.)
	Branch Circuit Concealed in Floor or Ceiling Below
	Low-Voltage Cable
5	Indicates Type of Heater (see schedule)
	Indicates Homerun to Panelboard (number of arrow heads indicates number of circuits)
WP	Weatherproof
M ½	Motor Outlet (numeral indicates horsepower)
J	Junction Box
D	Dimmer Control for Lighting Fixture
	Electric Baseboard Heater
	Flush-Mounted Electric Floor Heater
	Ceiling Electric Panel Heater
	Infrared Electric Heater, Ceiling Mounted
T	Double-Pole Thermostat for Electric Heat
F	Fire-Alarm Striking Station
G	Fire-Alarm Gong
D	Fire Detector
SD	Smoke Detector
B	Program Bell
Y	Yard Gong
M	Microphone, Wall-Mounted
	Microphone, Floor-Mounted
S	Speaker, Wall-Mounted
S	Speaker, Recessed

Figure 3-12 (cont.)

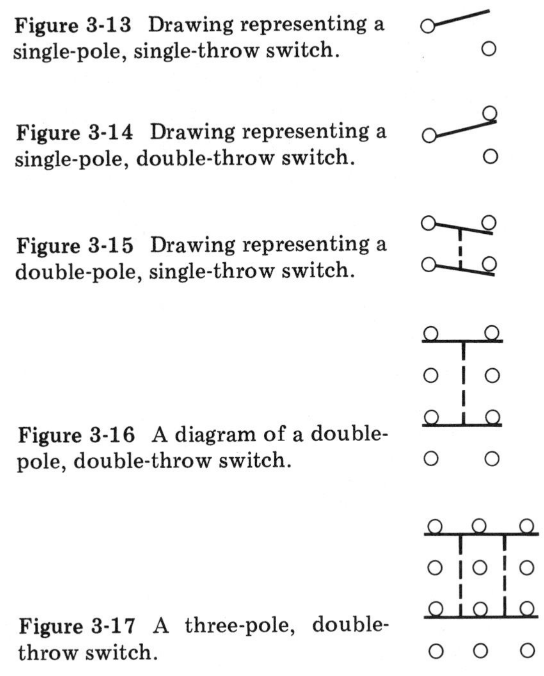

Figure 3-13 Drawing representing a single-pole, single-throw switch.

Figure 3-14 Drawing representing a single-pole, double-throw switch.

Figure 3-15 Drawing representing a double-pole, single-throw switch.

Figure 3-16 A diagram of a double-pole, double-throw switch.

Figure 3-17 A three-pole, double-throw switch.

Some examples of control symbols and abbreviations are shown in the illustrations to follow. For example, the drawing in Fig. 3-13 represents a single-pole, single-throw (SPST) switch; the contact on the left is normally open, while the one on the right is normally closed. The circuit in Fig. 3-14 represents a single-pole, double-throw (SPDT) switch, while the one in Fig. 3-15 shows double-pole, single-throw (DPST) contacts—one group of the normally open type and the other of the normally closed type. A double-pole, double-throw (DPDT) circuit is shown in Fig. 3-16, while a three-pole, double-throw (3PDT) circuit is represented in Fig. 3-17.

SCHEMATIC WIRING DIAGRAMS

Schematic wiring diagrams represent components in the control system by symbols, the wiring and connection to each, and other detail information. Sometimes the conductors are shown in an assembly of several wires, which appear as one line on the drawing. When this method is used, each wire should be numbered where it enters the assembly and should keep the same number when it comes out of the assembly to be connected to some component in the system. When reading or using such drawings, if the schematic does not follow this procedure, mark and number the wires yourself.

Although the symbols represent certain components, an exact description of each is usually listed in schedules or else noted on the drawings. Such drawings are seldom, if ever, drawn to scale as an architectural or cabinet drawing would be. They appear in diagrammatic form. In the better drawings, however, the components are arranged in neat and logical sequence so that they are easily traced and can be readily understood.

Electronic schematic diagrams indicate the scheme of plan according to which electronic or control components are connected for a specific purpose. Diagrams are not normally drawn to scale, and the symbols rarely look exactly like the component. Lines joining the symbols representing electronic or control components indicate that the components are connected.

To serve all its intended purposes, the schematic diagram must be accurate. Also, it must be understood by all qualified personnel, and it must provide definite information without ambiguity.

The schematics for a control circuit should indicate all circuits in the device. If they are accurate and well prepared, it will be easy to read and follow an entire closed path in each circuit. If there are interconnections, they will be clearly indicated.

In nearly all cases the conductors connecting the electronic symbols will be drawn either horizontally or vertically. Rarely are they ever slanted.

A dot at the junction of two crossing wires means a connection between the two wires. An absence of a dot in most cases indicates that the wires cross without connecting.

Schematic diagrams are, in effect, shorthand explanations of the manner in which an electronic circuit or group of circuits operates. They make extensive use of symbols and abbreviations. The more commonly used symbols were explained earlier in this chapter. These symbols must be learned to be able to interpret control drawings with the necessary speed required in the field or design department. The use of symbols presumes that the person reading the diagram is reasonably familiar with the operation of the device, and that he or she will be able to assign the correct meaning to the symbols. If the symbols are unusual, a legend will normally be provided to clarify matters.

Every component on a complete schematic diagram usually has a number to identify it. Supplementary data about such parts are supplied on the diagram or on an accompanying list in the form of a schedule, which describes the component in detail or refers to a common catalog number familiar in the trade.

To interpret schematic diagrams, remember that each circuit must be complete in itself. Each component should be in a closed loop connected by conductors to a source of electric current such as a transformer or line voltage. There will always be a conducting path leading from the source to the component and a return path leading from the component to the source. The path may consist of one or more conductors. Other components may also be in the same loop or in additional loops branching off to other devices. For each electronic component, it must be possible to trace a completed conducting loop to the source.

ELECTRICAL WIRING DIAGRAMS

Complete schematic electrical wiring diagrams used in highly complex heating control circuits are also represented by symbols. Every wire is either shown by itself or included in an assembly of several wires that appear as one line on the drawing. Figure 3-18 shows a complete schematic wiring diagram for a three-phase, ac magnetic nonreversing motor starter.

Note that this diagram shows the various devices in symbol form and indicates the actual connections of all wires between the devices. The three-wire supply lines are indicated by L_1, L_2, and L_3. The motor terminals of motor M are indicated by T_1, T_2, and T_3. Each line has a thermal overload-protection device (OL) connected in series with normally open line contactors C_1, C_2, and C_3, which are controlled by the

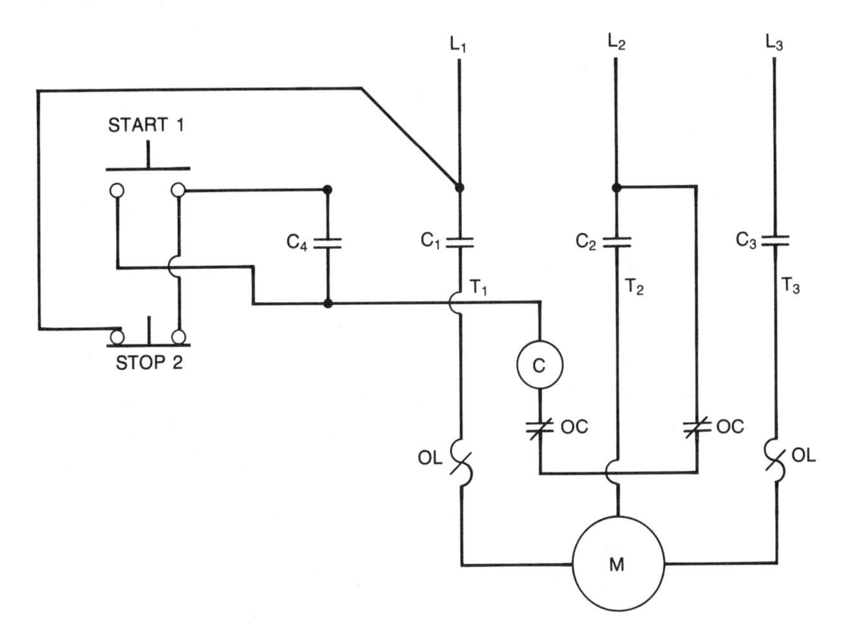

Figure 3-18 Schematic wiring diagram for a three-phase, ac magnetic nonreversing motor starter.

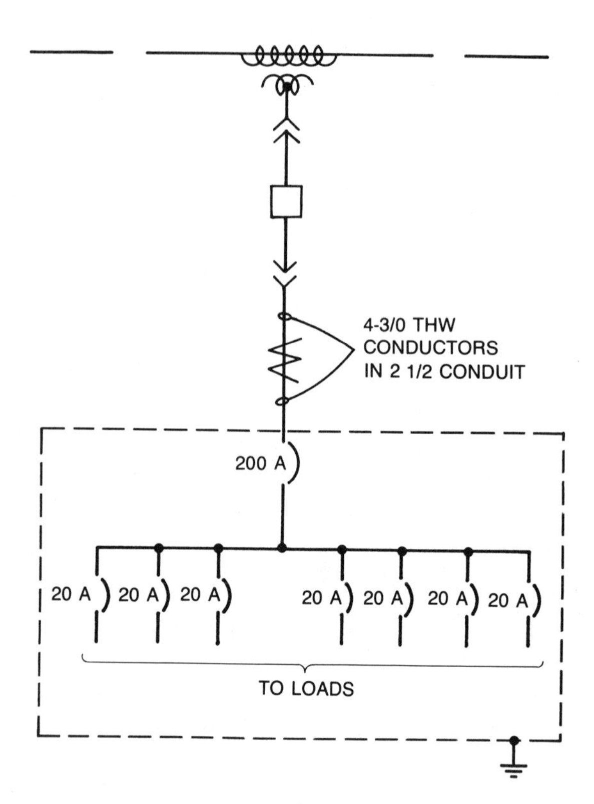

Figure 3-19 Typical single-line diagram of an industrial power distribution system.

magnetic starter coil, C. Each contactor has a pair of contacts that close or open during operation. The control station, consisting of start push button 1 and stop push button 2, is connected across lines L_1, and L_2. An auxiliary contactor (C_4) is connected in series with the stop push button and in parallel with the start push button. The control circuit also has normally closed overload contactors (OC) connected in series with the magnetic starter coil (C).

Figure 3-19 shows a typical single-line diagram of an industrial power distribution system. In analyzing this diagram, the utility company will bring its lines to a substation outside the plant building. Air switches, lightning arresters, single-throw switches, and an oil circuit breaker are provided there. This substation also reduces the primary voltage to 4160 V by transformers. Again, lightning arresters and other various disconnecting means are shown.

4

Tools

INTRODUCTION

High-quality work in any trade or profession can be accomplished only by the correct use of high-quality tools; the electrical industry is no exception. Any top-notch electrician or contractor will purchase only high-quality tools, and these will be kept in the best condition and will be replaced as needed.

Hand Tools

The hand tools commonly used by the electrician as set forth in the majority of labor agreements are as follows (1 of each except for channel locks where 2 pairs are normally supplied):

- Tool box
- Channel locks
- 6″ Screwdriver
- 6′ Folding rule
- 2″ Conduit reamer
- Expansion bit
- Combination square
- 1/2″ Chisel
- 1/2″ Wood chisel
- 8″ Side cutting plier
- 10″ Screwdriver
- Claw hammer
- Voltage tester
- Brace
- Bit extension
- Center punch
- Tap wrench
- 6″ Crescent wrench

10″ Crescent wrench	50′ Steel tape
Keyhole saw	Phillip's screwdriver
Gripping screwdriver	10″ Tin snips
Small architect's scale	Electrician's knife
8″ Level	Diagonal pliers
Long nose pliers	Lock
Hacksaw frame	10″ Mill file
Fuse puller	Flashlight (preferably wired for testing continuity)

This list of hand tools is the minimum essential hand tools needed in order to perform good work on the usual kind of electrical installations. Any other tools needed are normally supplied by the contractor.

Knowing the Tools

As with other types of electrical materials, it is also necessary for all personnel to have a good knowledge of the various kinds of tools and installation equipment that are necessary to perform various electrical installations.

While workers with any amount of experience at all are usually adept at using the installation tools and equipment of the trade, there may be instances where workers may have acquired bad habits or awkward movements, which, if corrected, increase their productive efficiency. For example, bad habits common to many workers include

1. Not cleaning out threading dies or keeping them tight in the stock,
2. Continuing to use dull dies and hack saw blades,
3. Not using sufficient cutting fluid while cutting threads,
4. Throwing diestocks and hack saws to the deck or floor instead of leaning or hanging them on the vise stand,
5. Assuming an awkward stance when turning a diestock,
6. Putting too much pressure on a hack saw,
7. Pulling wire with a tangle of fish-steel about their feet,
8. Using a screw driver with too small or too large a bit,
9. Failing to properly adjust crescent type wrenches or using ill-fitting end wrenches with resultant damage to boltheads and injured knuckles when the wrench slips;
10. Failing to properly adjust pipe wrenches or tongs resulting in "chewed-up" conduit and possible injury to the worker if the wrench or tong slips under a heavy pull,

11. Using tools for the wrong purpose, for example, using a pair of pliers and a screw driver as a hammer and punch.

Anyone involved with electrical construction should be alert to such bad habits and awkward movements.

Many new and improved hand tools and power operated tools are being introduced into the industry, which can be used to increase productivity, lessen fatigue and reduce injuries. Many otherwise experienced workers or contractors may not be familiar with the application and uses of some of these items.

Proper Care and Use of Tools

The following is a brief summary of tool use and maintenance; other techniques will follow:

1. Sufficient space should be provided for, to protect all tools, equipment, and clothing.
2. Company tools should be used only for the purpose for which they are intended.
3. Chisels, star drills, drill, and all cutting tools are to be kept sharp. All impact tools should be kept dressed to prevent "mushrooming."
4. The right size and proper type wrenches must be used.
5. No power tools should be hung by hose or cord.
6. All work should be conducted in a safe manner and all equipment should be so maintained, handled, and stored so as to avoid any danger to persons employed on the job.
7. Safety equipment should be handled and stored in the proper places and equipment such as rubber gloves and blankets should be electrically tested and tagged at least every two months and should be air tested each time before they are used.
8. Don't throw tools or material from scaffolds, platforms, or walkways. Use hand lines to lower such items.

In any electrical business, certain personnel are usually designated to be responsible for the maintenance and replacement of the tools and equipment so they will always be in satisfactory condition for use on the job to avoid lost time of workers in attempting to perform their work with badly worn or broken tools or equipment that are not in proper operating condition.

Electrical contractors cannot expect workers to perform satisfactory work in a normal length of time when they are provided with tools and equipment in an unsatisfactory or inoperative condition. The psychological reaction is also bad, as even first-class workers will take the attitude that if the contractor does not take the pains to provide them with satisfactory tools, why should they be concerned about performing the work in a workmanlike manner and in the minimum amount of time.

Replacement of Tools and Equipment

When the condition of tools and equipment becomes such that they cannot be restored to satisfactory condition by maintenance or repairing, or they become obsolete, they should be replaced immediately. Retention of such items in the tool and equipment stock gives a false impression as to what items are on hand, and there is always a possibility that they will be shipped to a job for one last period of use. This is really false economy if one considers the possibility of lost time of workers attempting to use them or possible injury or loss of life.

Many contractors will make an effort to obtain the most modern and adaptable tools and equipment which will "stand up" the longest and will require the least maintenance and repair. They know that the purchase of an inferior item because its initial cost may be slightly lower than a better item is false economy.

General Classification of Tools

1. Shop tools and equipment
2. Hand tools—small and medium sizes
3. Expendable tools
4. Conduit and other raceway tools
5. Portable power tools
6. Powder actuated tools
7. Wire and cable installation tools
8. Underground installation tools
9. Cable splicing tools
10. Wearing apparel
11. Mechanical trenching and digging equipment
12. Line construction tools and equipment
13. Trucks and other automotive equipment
14. Safety equipment

15. Testing and measuring equipment
16. Special facilities

Certain items of tools and equipment are more generally confined to and used in the shop to process, fabricate, or test electrical materials, fixtures, operating equipment and the like. In most cases, most such work may be performed before the materials are delivered to the job site, or in some instances, material may be returned from the job to the shop, the necessary work performed, and the item then returned to the job.

When the size or type of the job is large enough, a job-site shop may be set up equipped, to at least some extent, with shop-type tools.

The size of the contractor's operations may be a determining factor as to the type of tools and equipment kept in the shop rather than being placed on the jobs. For example, a relatively small operation might maintain a power hack saw and, say a pipe threading machine in the shop only, whereas a larger contractor would not only maintain such equipment in the shop but find it advisable to also place such items of equipment on certain large jobs.

While it is not possible to draw an absolute line of distinction between items of tools and equipment that are maintained in a shop or used on jobs, the following list is representative of those maintained in a well-equipped shop.

Work benches	Drill press
Electric tool grinder	Power hack saw
Power threader	Hydraulic bender
Lathe	Heavy metal shear and punch
Electric welder and equipment	Forge
Anvil	Air compressor
Machinist vise	Pipe vise
Set of taps	Set of socket wrenches
Set of rod stock and dies	Hand truck
Dollies	Power threading tools
Hand and power bending tools	Conduit cutting devices
Fish tapes and vacuum fish tape machines	Set of steel lettering and numbering dies
Oxyacetyline tank, gauges, and burning equipment	Wire and cable reeling and measuring equipment

The use of certain items of tools and equipment is not confined to the performance of any particular phase of electrical construction. At least some such items are usually required on any job. These items include the following:

- Padlocks
- Extension ladders
- Chain hoists
- Dollies
- Tarpaulins
- Walkie-talkie
- Tool and storage boxes, and/or tool shed
- Step ladders
- Rolling scaffolds
- Wagon trucks
- Blocks and falls
- Transit and level
- Extension or drop cords
- Ladder and equipment securing chains

SPECIFIC TYPES OF TOOLS

Conduit and Raceway Tools

A large proportion of the work on any electrical construction project involves the installation of various raceways and conduit systems. The tools required for this work include the following:

- Pipe vise and stand
- Stocks and dies

Powder-Actuated Fastening Systems

During this century concrete has revolutionized the construction industry and greatly affected the development of power tools.

The principle of the powder actuated tool is to fire a fastener into material and anchor or make it secure to another material. Some applications are wood to concrete, steel to concrete, wood to steel, steel to steel and numerous applications of fastening fixtures, electrical boxes, and so on to concrete or steel.

This type of system is very competitive with other fastening systems such as concrete anchors, Tek screws, and so on, and there are areas where it may or may not be preferred.

A powder-actuated tool in simple terms is a "pistol" which fires a bullet composed of two elements: a cartridge with firing cap and powder, and a bullet, which in this case is actually a fastening device. In use, the fastening device is inserted into the barrel, and then the correct cartridge is inserted (depending upon what material the device is being fired into). The tool is pressed against the work surface, the trigger is pulled, and the fastener then travels in free flight to its destination. Although this system is simple to use, there are precautions and safeguards that must be observed.

The tools are identified by types and classes. In general, there are two types in current use: direct-acting and indirect-acting. In the former type, it is a tool in which the expanding gas of a power load acts directly on the fastener to be driven into the work. The latter type is a tool in which the expanding gas of a power load acts on a captive piston which in turn drives the fastener into the work.

This classification may be broken down further into three classes; that is, low velocity, medium velocity, and high velocity—varying from 300 to 500 ft/s.

Fasteners. The fasteners used in powder-actuated tools are not common nails. They are manufactured from special steel and heat-treated to produce a very hard, yet ductile fastener. These properties are necessary to permit the fastener to penetrate concrete or steel without breaking.

The fastener is equipped with some type of tip, washer, eyelet, or other guide member. This guide aligns the fastener in the tool as it is being driven and is usually used to retain the fastener in the tool.

The two basic types of fasteners used are drive pins and threaded studs.

Drive pin: A special nail-like fastener designed to permanently attach one material to another such as wood to concrete or steel. Head diameters are generally 1/4 in., 5/16 in., or 3/8 in. However, for additional head bearing in conjunction with soft materials, washers of various diameters are either fastened through or made a part of the drive pin assembly.

Threaded stud: A fastener composed of a shank portion which is driven into the base material and a threaded portion to which an object can be attached with a nut. Usually thread sizes are 8-32, 10-24, 1/4-20, 5/16-18, and 3/8-16.

There are also other types of special fasteners designed for specific applications.

Eye pin: A fastener with a hole through which wires, chains, etc. can be passed for items such as hanging ceilings and light fixtures.

Utility stud: A threaded stud with a threaded collar, which can be tightened or removed after the fastener has been driven into the work surface.

Power loads. The power load is a unique, portable, self-contained energy source in powder-actuated tools. These power loads are available in two forms: cased and caseless. The propellant in a cased power load is contained in a metallic case.

Cased power loads are available in various sizes ranging from .22 through .38 caliber, and caseless loads are available in various sizes and shapes.

The caseless power load does not have a case and the propellant is in a solid form.

Regardless of the type, caliber, size, or shape, there is a standard number and color code used to identify the power level or strength of all power loads.

Cased power loads used in all types and classes of tools cover a range of 12 power load levels numbered 1 through 12, with the lightest being No. 1 load and the heaviest being No. 12 load. A basic six-color code of gray, brown, green, yellow, red, and purple is used twice because there are not 12 different readily distinguishable permanent colors. Power loads No. 1 through 6 are in brass colored cases and power loads No. 7 through No. 12 are in nickel colored cases. It is the combination of the case color and load color that defines the load level or strength. Each cased power load is clearly identified on one end by its power level color.

The following chart shows this simple number and color identification code.

	Color Identification	
Power Level	Case Color	Load Color
1	Brass	Gray
2	Brass	Brown
3	Brass	Green
4	Brass	Yellow
5	Brass	Red
6	Brass	Purple
7	Nickel	Gray
8	Nickel	Brown
9	Nickel	Green
10	Nickel	Yellow
11	Nickel	Red
12	Nickel	Purple

Caseless loads are manufactured only in the No. 1 through No. 6 load levels and are color coded with the basic load colors gray, brown, green, yellow, red, and purple.

In addition to the identification of the power load, each package is color coded and shows the load level number.

Power load selection. In selecting the proper power load to use for any application, it is important to start with the lighest power load level recommended for the tool being used. If in using the lightest load the first test fastener does not penetrate to the desired depth, the next higher power load should be tried. If necessary, continue increasing power levels by single steps until proper penetration is obtained.

There are numerous special tool accessories available for use with all tools. Examples of these are adapters that hold various types of clips, brackets, or washers at the muzzle end of the tool.

Base materials. The material into which the fastener shank is driven and from which holding power is obtained is known as the base material.

In general, base materials are metal and masonry of various types and hardness. The majority of base materials found on the job are suitable for powder actuated fastening. However, it is very important that qualified operators be able to determine the suitability of any material into which they intend to drive a fastener.

When pierced by the fastener, suitable base materials will expand or compress, or both, and have sufficient hardness and thickness to produce holding power and not allow the fastener to pass completely through.

Unsuitable base materials can be put into three categories:

1. *Too hard*—Fastener will not be able to penetrate and could possibly deflect or break. Examples are hardened steel, welds, cast steel, marble, spring steel, natural rock.
2. *Too brittle*—Material will crack or shatter and fastener could deflect or pass completely through. Examples are glass, glazed tile, brick, slate.
3. *Too soft*—Material does not have the characteristics to produce holding power and fastener could pass completely through. Examples are wood, plaster drywall, composition board, plywood.

Masonry materials suitable for fastening consist of the following:

1. Poured concrete
2. Precast concrete
3. Prestressed concrete
4. Concrete block
5. Mortar joints (horizontal)

It is important to understand what happens when a fastener is driven into any masonry material and why the fastener holds. The holding power of the fastener results primarily from a compression bond of the masonry to the fastener shank. The fastener, on penetration, displaces the masonry which tries to return to its original form and exerts a squeezing effect.

Compression of the masonry around the fastener shank takes place with the amount of compression increasing in relation to the depth of penetration and the compressive strength of the masonry.

When the depth of penetration produces a bond on the fastener shank equal to the strength of the masonry, the maximum holding power results.

It is important to understand that soft masonry has a low compressive strength [2000 pounds per square inch (psi)] and hard masonry has a high compressive strength (5000 psi or more).

Harder masonry results in greater material compression giving a higher bond with greater holding power at the same fastener penetration.

One should not fasten closer than 3 in. from the edge of masonry. If the masonry cracks, the fastener won't hold and there's a chance a chunk of masonry or the fastener could break off and hurt someone.

Setting fasteners too close together can also cause masonry to crack. Recommended minimum fastener spacing based on shank diameter is as follows:

Shank Diameter (in.)	Recommended Minimum Distance Between Fastenings (in.)
1/8 thru 5/32	3
11/64 thru 3/16	4
7/32 thru 1/4	6

DRILL MOTOR

The drill motor is the backbone of the portable electric tool industry. It takes on many forms, from drills to hammer-drills to diamond drills to magnetic drills, but the *pistol drill* or portable hand drill is the most common.

As the industry progressed and pistol-type tools were designed, it quickly became popular for its compactness, light weight, and ease of handling in drilling small holes. Today, when an electric tool is mentioned, the first response is a 1/4-in. electric drill. However, many other, and larger, types are available for use in the electrical industry.

As you review the various types of drill motors, you will note that they run a large range of revolutions per minute (rpm), and also carry a range of such capacities from 1/4 in. through 1/2 in. What is not commonly understood is how these ratings are determined. Contrary to popular belief, the motor itself has nothing to do with the capacity rating of a drill motor. There is a table of standards that informs us that, say, a 1/4-in. high-speed drill bit can turn a 2000-2500 rpm under continuous operation in steel plate and will not damage the bit; that is, the bit will not become damaged or turn blue from excessive speed. The same applies to a 3/8-in. high-speed bit turning at 650-1000 rpm. There is no reference to the motor since it is not determined whether the motor is capable of withstanding a continuous load of drilling in those capacities in steel. From this, it is easy to understand why an inexpensive drill can obtain the same capacity rating of a quality industrial rated drill.

Since all quality tools are designed for continuous operation within their rated capacity, you can quickly determine that a 1/4-in. drill motor should use a maximum high-speed bit of 1/4 in. in steel, or a 3/8-in. bit in a 3/8-in. drill motor. This same rule of thumb applies to all tools throughout the line. Capacity in hardwood is usually twice that in steel.

VACUUM CLEANERS

Next to the common drill motor, probably nothing in the power tool line reaches a broader market than the vacuum. It is a very common item as one can be found in nearly every household in this country. In general, it is a high-speed motor that creates suction to lift and remove dirt, dust, and debris from floors, rugs, drapes, shelves, and machinery, as well as an endless list of other uses.

The applications for industrial vacuums challenge the imagination: for cleanup work, cleaning controls, removing lint, cleaning dust from drilled holes, for fish tape systems, and many other uses.

Two types of models are in common use: the dry pickup and the wet/dry models. Manufacturers' catalogs contain complete descriptions.

ELECTRIC HAMMERS

When man first attempted to penetrate concrete, it was necessary to obtain a star drill and heavy hammer. The star drill was hit once, turned slightly, and then hit again; this procedure was repeated until sufficient chips were removed to make a hole. Drilling a hole in this manner is hard work and as power tools were developed, a tool emerged that would do most of the work; namely, the electric hammer. It is still found in the industry for use in chipping, chiseling, scaling, cleaning of bricks, cleaning out motors, and so on.

When the common electric hammer was used, the operator had to turn the star drill as the hammer pounded. For even less effort on the operator's part, the rotary hammer was developed. These tools both hammered and rotated at high speeds, and were capable of drilling 1-1/2 in. with solid bits and 2-1/2 in. with core bits. Then the anchor manufacturers started producing anchors that would fill holes drilled by these rotary hammers. Exotic drop-in anchors required smaller holes, so there became a pressing need for smaller rotary hammers capable of drilling 3/4-in. diameter holes and smaller.

A hammer–drill is a simple dual-purpose tool, as its name implies. It is a high-speed hammer, and it is a drill. These small hammer–drills fell into the marketplace for much the same reasons that medium-size hammers did. Anchor manufacturers developed a large range of drop-in anchors, which created a demand for a small high-speed, hammer–drill.

PORTABLE BAND SAW

There are many times when it is impractical to take work to a large band saw, so it is more practical to take the saw to the work. Workers in the electrical industry have found many uses for the portable band saw: cutting conduit, hanger rods, extrusions, bolts, cable, plastics and other materials.

Band saws use two rotating wheels with rubber tires that drive a continuous blade. Several different types of blades are available for cutting various materials, but the most common are blades made from carbon steel, alloy steel, and high-speed steel. Blades are available with 6, 8, 10, 14, 18, and 24 teeth per inch. In cutting metals, the rule of thumb is to always have three teeth in the material at all times. Using too coarse a blade can cause thin metals to hang up in the gullet between two teeth, and tear out a section of teeth.

Choose a blade with as few teeth per inch as possible to accomplish the cutting quickly, but keep in mind the rule of three teeth in the cut.

MISCELLANEOUS POWER TOOLS

The electrical industry, probably more so than other trades, has developed a power tool for practically all operations: pulling wire and cable, blowing or sucking fish tapes, drilling holes, bending conduit, cutting conduit, threading conduit, pushing conduit under roads, and many, many others. Any electrical contractor with sufficient work can benefit from any of them—provided they know the availability of the tools and the proper method of using them.

5

Electrical Boxes

Outlet boxes normally fall into three categories: (1) pressed steel boxes with knockouts of various sizes for raceway or cable entrances, (2) cast iron, aluminum, or brass boxes with threaded hubs of various sizes and locations for raceway entrances, and (3) nonmetallic boxes.

Pressed steel boxes also fall into two categories, those with conduit, electric metallic tubing, and cable, and those designed for use with specific makes of surface metal moldings.

Outlet boxes vary in size and shape depending upon their use, the size of the raceway, the number of conductors entering the box, the type of building construction, atmospheric conditions of the area, and special requirements.

Outlet box covers usually are required to adapt the box to the particular use it is to serve. For example, a 4-in. square box is adapted to one-gang or two-gang switches or receptacles by the use of either one-gang or two-gang flush device covers, or a one-gang cast hub box is adapted to provide a vaporproof switch or a vaporproof receptacle cover.

Special outlet box hangers are available to facilitate their installation, particularly in frame building construction.

Electrical drawings rarely indicate the exact types and sizes of outlet boxes to be used for a given application, with the possible exception of boxes used in hazardous areas. Therefore, those involved in electrical construction must have a good knowledge of all types of outlet boxes and be able to select the best types (and the correct sizes) for any given application. For example, outlet boxes for use with type NM cable (nonmetallic sheath cable) should contain built-in cable clamps specifically designed for this type of cable.

The time involved during the installation of raceway or cable systems can be greatly reduced if the proper outlet boxes are used. Figure 5-1 illustrates several types of outlet boxes currently available. Those unfamiliar with the numerous types of other outlet boxes should

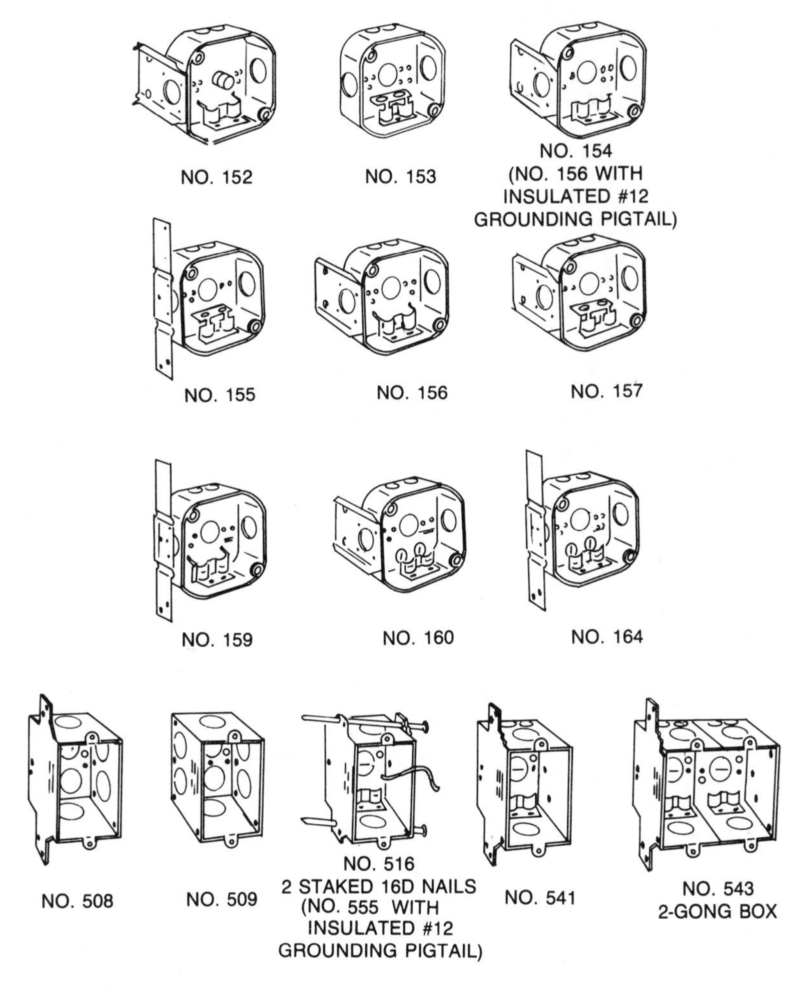

Figure 5-1 Several types of outlets boxes currently available for use in wiring systems.

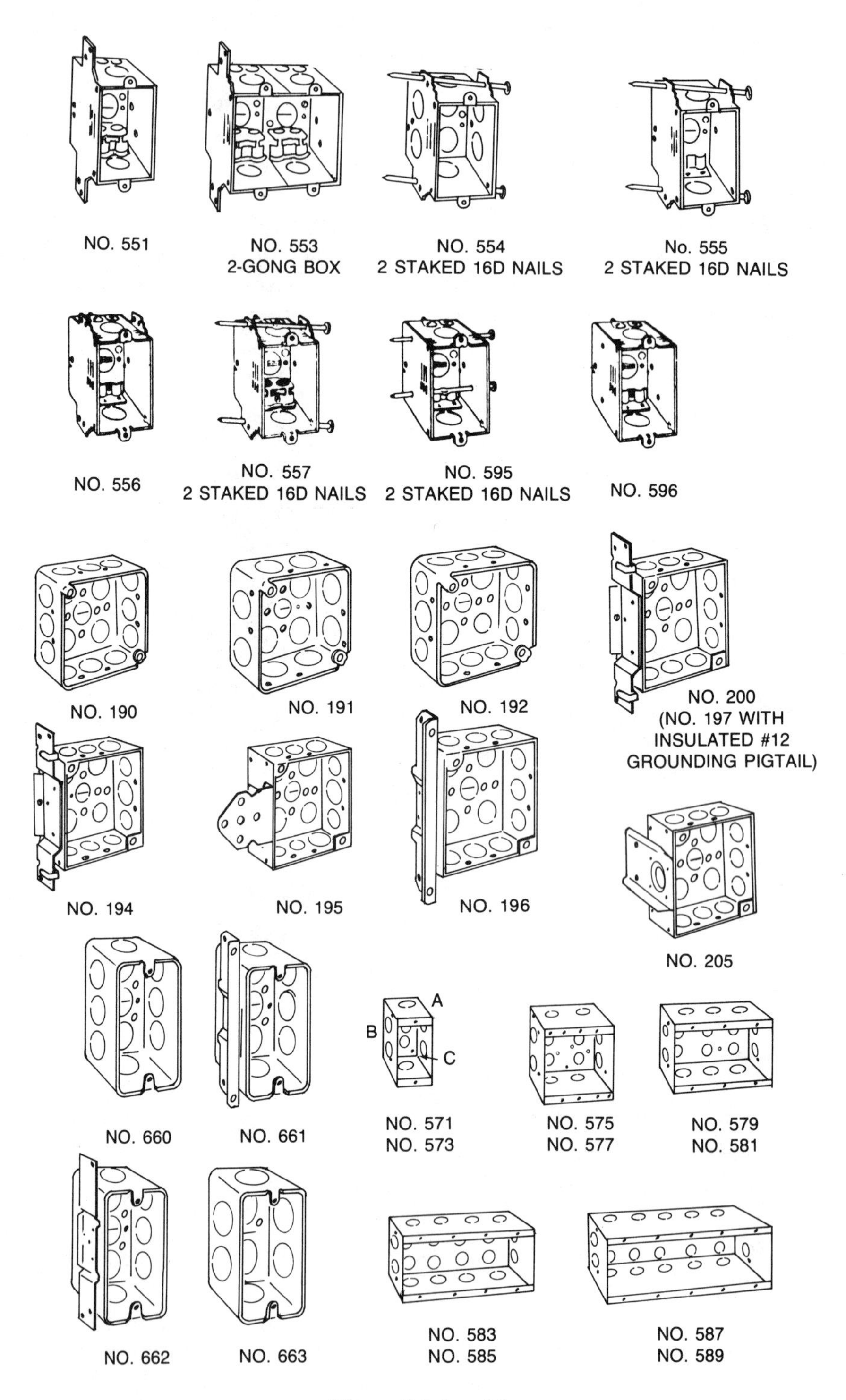

Figure 5-1 (cont.)

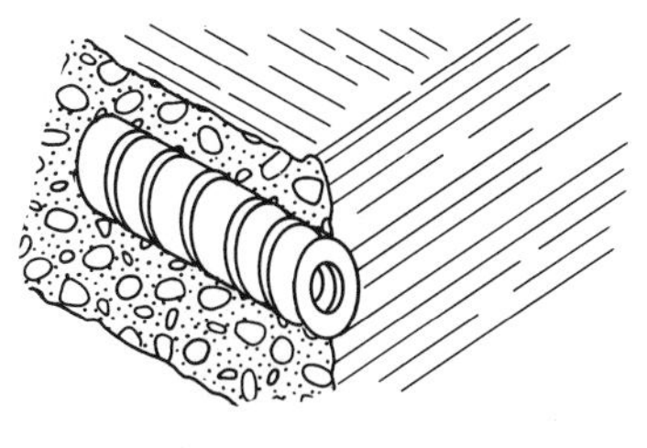

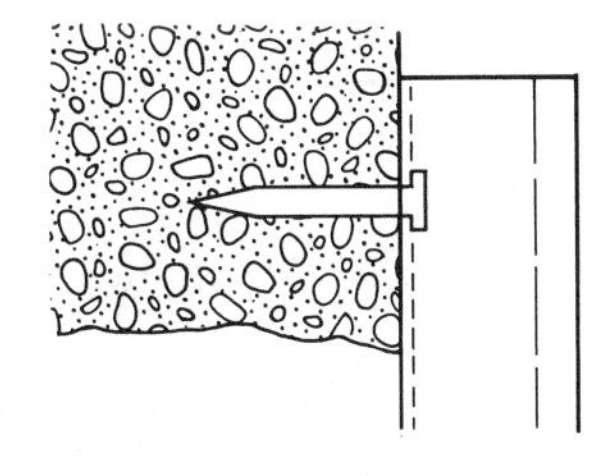

Figure 5-2 Outlet boxes can be secured to masonry walls by using masonry anchors, or powder-actuated fasteners.

obtain current catalogs from the outlet box manufacturers and study the illustrations and descriptions. With this basic information, on-the-job installation will go smoother.

Outlet boxes must be securely fastened in place, unless otherwise provided for specific purposes in the NE Code. Boxes attached to wood studs are normally secured with nails; boxes attached to metal studs are secured with sheet-metal screws; boxes attached to metal beams are normally secured with nuts and bolts or welded in place; boxes attached to masonry walls are secured with masonry expansion anchors or powder-actuated fasteners (Fig. 5-2), and so forth.

Article 370-13 of the NE Code requires the following procedures for mounting outlet boxes.

> Boxes shall be securely and rigidly fastened to the surface upon which they are mounted or securely and rigidly embedded in concrete or masonry. Where nails are used as a mounting means and pass through the interior of the box, they shall not be more than 1/4 inch from the back of the box. Boxes shall be supported from a structural member of the building either directly, by using a substantial and approved metal or wooden brace, or as otherwise provided in this section. If made of wood, the brace shall not be less than nominal one-inch thickness. If made of metal, it shall be corrosion resistant and not be less than No. 24 MSG.
>
> Where mounted in new walls in which no structural members are provided or in existing walls in previously occupied buildings, boxes not over 100 cubic inches in size, specifically approved for the purpose shall be affixed with approved anchors or clamps so as to provide a rigid and secure installation. Threaded boxes or fittings not over 100 cubic inches in size that do not contain devices or support fixtures shall be considered adequately supported if two or more conduits are threaded into the box wrench tight and are supported within three feet of the box on two or more sides, as is required by this section. Threaded boxes or fittings not over 100 cubic inches in size shall be considered to be adequately supported if two or more conduits are threaded into the box wrench tight and are supported within 18 inches of the box, as required by this section.

BOX DIMENSIONS (IN.) TRADE SIZE	CAPACITY (IN.^{3})	MAXIMUM NUMBER OF CONDUCTORS			
		NO. 14	NO. 12	NO. 10	No. 8
3 1/4 × 1 1/2 OCTAGONAL	10.9	5	4	4	3
3 1/2 × 1 1/2 OCTAGONAL	11.9	5	5	4	3
4 × 1 1/2 OCTAGONAL	17.1	8	7	6	5
4 × 2 1/8 OCTAGONAL	23.6	11	10	9	7
4 × 1 1/2 SQUARE	22.6	11	10	9	7
4 × 2 1/8 SQUARE	31.9	15	14	12	10
4 11/16 × 1 1/2 SQUARE	32.2	16	14	12	10
4 11/16 × 2 1/8 SQUARE	46.4	23	20	18	15
3 × 2 × 1 1/2 DEVICE	7.9	3	3	3	2
3 × 2 × 2 DEVICE	10.7	5	4	4	3
3 × 2 × 2 1/4 DEVICE	11.3	5	5	4	3
3 × 2 × 2 1/2 DEVICE	13.0	6	5	5	4
3 × 2 × 2 3/4 DEVICE	14.6	7	6	5	4
3 × 2 × 3 1/2 DEVICE	18.3	9	8	7	6
4 × 2 1/8 × 1 1/2 DEVICE	11.1	5	4	4	3
4 × 2 1/8 × 1 7/8 DEVICE	13.9	6	6	5	4
4 × 2 1/8 × 2 1/8 DEVICE	15.6	7	6	6	5

Figure 5-3 Maximum number of conductors that may terminate in various sizes of outlet boxes.

SIZING OUTLET BOXES

The NE Code lists specific instructions for sizing outlet boxes for the number and sizes of conductors entering the box. Figure 5-3, for example, gives the maximum number of conductors that may terminate in various sizes of outlet boxes. This table applies where no fittings or devices, such as cable clamps, switches, and receptacles, are contained in the box and where no grounding conductors are part of the wiring within the box. When one or more cable clamps or fixture studs are contained in the box, the number of conductors given in the table must be reduced.

The number of conductors given in the table must also be deducted for each switch, receptacle, or other wiring device attached to the box. Further deductions must be made for one or more grounding conductors.

BRANCH CIRCUIT WIRING BOXES

The branch circuit wiring extending from the panelboards to the various outlets is not confined to any particular part of the building structure and is placed in the floors, walls, partitions and ceilings, exposed and concealed. The outlet boxes are fastened to the building structure when the wiring is exposed. When the wiring is concealed, the outlet boxes are placed in the building structures and supported by wooden backing, metal straps, or by the concrete when the wiring system is in a concrete slab. In concealed wiring, the depth at which the fronts of the boxes are placed with respect to the rough building structure depends upon the final building structure ceiling and wall finish and the type and depth of outlet box covers used.

In the case of reinforced concrete construction, the boxes for ceiling outlets for the floor below are secured to the horizontal forms or deck with the normal opening facing down. Special boxes which have a removable back plate are ordinarily used.

Outlet boxes for floor outlets are put in place on the deck, with the box opening facing upward and at a proper height to allow the finished outlet to be flush with the finished floor surface. Metal stirrups are often used to securely hold the boxes in place at the proper height during the pour. The conduits are put in place after the steel reinforcing bars or the bottom tiers of such bars are put in place and are usually tied down to the reinforcing steel with iron wire. Metal stirrups or saddles are used to support floor outlet boxes securely at the proper height.

While outlet boxes are often used as junction or pull boxes, the industry has developed a wide variety of hub fittings, both threaded and threadless, for rigid conduit and electric metallic tubing for all sizes of raceway. These fittings provide for right angle change in direction (e.g., LB, LL, LR), a right angle tap (e.g., T, TB), a right angle cross (X and XA) and many special adaptations.

OUTLET BOXES IN HAZARDOUS LOCATIONS

Any area in which the atmosphere or a material in the area is such that the arcing of operating electrical contacts, components, and equipment may cause an explosion or fire is considered a hazardous location. In all such cases, explosion-proof equipment, raceways, and fittings are used to provide an explosion-proof wiring system, including the outlet boxes.

Hazardous locations have been classified in the NE Code into certain class locations. Various atmospheric groups have been established on the basis of the explosive character of the atmosphere for the testing and approval of equipment for use in the various groups.

Class I Locations: Those locations in which flammable gases or vapors may be present in the air in quantities sufficient to produce explosive or ignitible mixtures are classified as Class I locations. Examples of such locations are interiors of spray paint booths where volatile, flammable solvents are used, inadequately ventilated pump rooms where flammable gas is pumped, and drying rooms for the evaporation of flammable solvents.

Class II Locations: Class II locations are those that are hazardous because of the presence of combustible dust. Class II, Division 1 locations are areas where combustible dust, under normal operating conditions, may be present in the air in quantities sufficient to produce explosive or ignitible mixtures; examples are working areas of grain handling and storage plants and rooms containing grinders or pulverizers. Class II, Division 2 locations are areas where dangerous concentrations of suspended dust are not likely, but where dust accumulations might form.

Class III Locations: These locations are those areas that are hazardous because of the presence of easily ignitible fibers or flyings, but such fibers and flyings are not likely to be in suspension in the air in these locations in quantities sufficient to produce ignitible mixtures. Such locations usually include some parts of rayon, cotton, and textile mills; clothing manufacturing plants; and woodworking plants.

The wide assortment of explosion-proof equipment now available makes it possible to provide adequate electrical installations under any of these hazardous conditions. However, the electrician must be thoroughly familiar with all NE Code requirements, and know what fittings are available, how to install them properly, and where and when to use the various fittings.

The usual working drawings for a hazardous area are drawn the same as the layout of an electrical system for a nonhazardous area—the only distinction is a note on the drawings stating that the wiring in this particular area shall conform to the NE Code requirements for hazardous locations. The designer will sometimes add the letters *EXP* next to all the symbols of the outlet that are to be explosion proof. However few engineers or draftsmen detail their drawings for hazardous areas sufficiently for the electricians to proceed with the installation without additional study and layout work on the job site. Therefore, an electrician must be familiar with the layout and installation procedures before attempting such an installation. For other than very simple installations, it may be advisable to make rough, detailed wiring layouts of the proposed installation, even if they are merely sketches on the original working drawings.

When an electrical system in a hazardous location is designed or installed, the type of building structure and finish must be considered. If the building is under construction, this information may be obtained from the architectural drawings and specifications. If the installation is made in an existing building, a preliminary job-site investigation is often necessary. The location of the explosion-proof outlets, whether concealed or exposed, and the class of hazardous locations should appear in the electrical drawings and specifications. If such information is not provided, the contractor or electrician will have to determine this information—from the architect, owner, or local inspection authority.

In general, rigid metallic conduit is required for all hazardous locations, except for the special flexible terminations and as otherwise permitted in the NE Code. The conduit should be threaded with a standard conduit cutting die that provides 3/4-in. taper per foot. The conduit should be made up wrench tight in order to minimize sparking in the event fault current flows through the conduit system (NE Code Article 500-1). Where it is impractical to make a threaded joint tight, a bonding jumper should be used. All boxes, fittings, and joints shall be threaded for connection to the conduit system and shall be an approved, explosion-proof type. Threaded joints shall be made up with at least five threads fully engaged. Where it becomes necessary to employ flexible connectors at motor or fixture terminals, flexible fittings approved for the particular class location shall be used.

Seal-off fittings are required in conduit systems to prevent the passage of gases, vapors, or flames from one portion of the electrical installation to another through the circuit. For Class I, Division 1 locations, the NE Code (Article 501-5) states that there shall be seal-off fittings

> In each conduit run entering an enclosure for switches, circuit breakers, fuses, relays, resistors or other apparatus which may produce arcs, sparks or high temperatures, seals shall be placed as close as practicable and in no case more than 18 inches from such enclosures. There shall be no junction box or similar enclosure in the conduit run between the sealing fitting and the apparatus enclosure . . .
>
> In each conduit run of 2-inch size or larger entering the enclosure or fitting housing terminals, splices or taps, and within 18 inches of such enclosure or fitting . . .
>
> In each conduit run leaving the Class I, Division 1 hazardous area, the sealing fitting may be located on either side of the boundary of such hazardous area, but shall be so designed and installed that any gases or vapors which may enter the conduit system, within the Division 1 hazardous area, will not enter or be communicated to the conduit beyond the seal. There shall be no union, coupling, box or fitting in the conduit between the sealing fitting and the point at which the conduit leaves the Division 1 area . . .

Sealing compound shall be approved for the purpose, shall not be affected by the surrounding atmosphere or liquids, and shall not have a melting point of less than 200°F (93°C). Most sealing-compound kits contain a powder in a polyethylene bag within an outer container. To

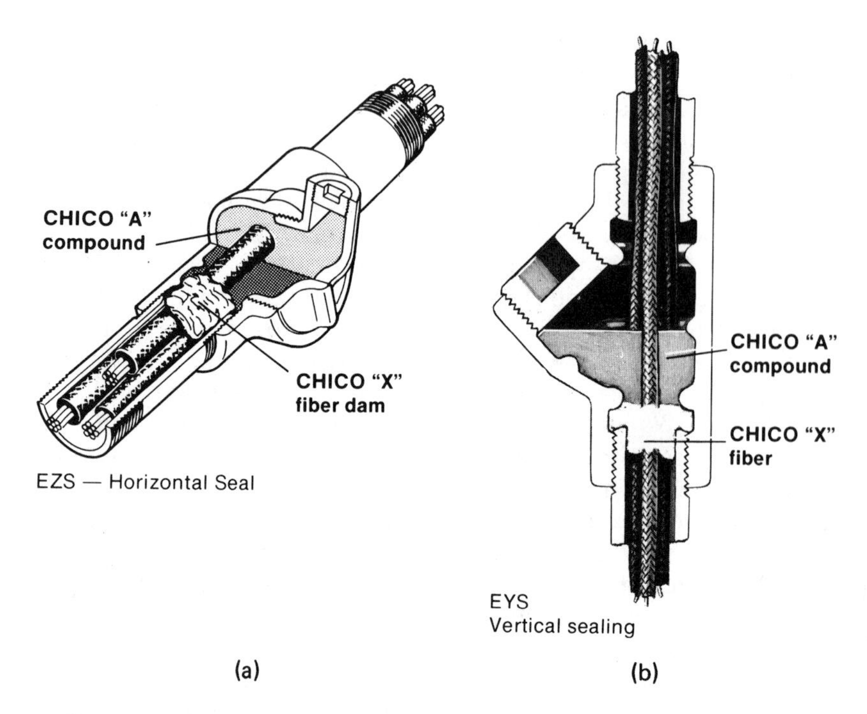

Figure 5-4 Types of seal-off fittings in use. (Courtesy of Crouse-Hinds.)

mix, remove the bag of powder, fill the outside container with water up to the marked line on the container, and pour in the powder and mix.

To pack the seal off, remove the threaded plug or plugs from the fitting and insert the asbestos fiber supplied with the packing kit. Tamp the fiber between the wires and the hub before pouring the sealing compound into the fitting. Then pour in the sealing cement and reset the threaded plug tightly. The fiber packing prevents the sealing compound (in the liquid state) from entering the conduit lines.

The seal-off fittings in Fig. 5-4 are typical of those used. The type in Fig. 5-4(a) is for vertical mounting and is provided with a threaded, plugged opening into which the sealing cement is poured. The seal-off in Fig. 5-4(b) has an additional plugged opening in the lower hub to facilitate packing fiber around the conductors in order to form a dam for the sealing cement.

Most other explosion-proof fittings are provided with threaded hubs for securing the conduit. Typical fittings include switch and junction boxes, conduit bodies, union end connectors, flexible couplings, explosion-proof lighting fixtures, receptacles, and panelboard and motor starter enclosures. A practical representation of these and other fittings is shown in Fig. 5-5.

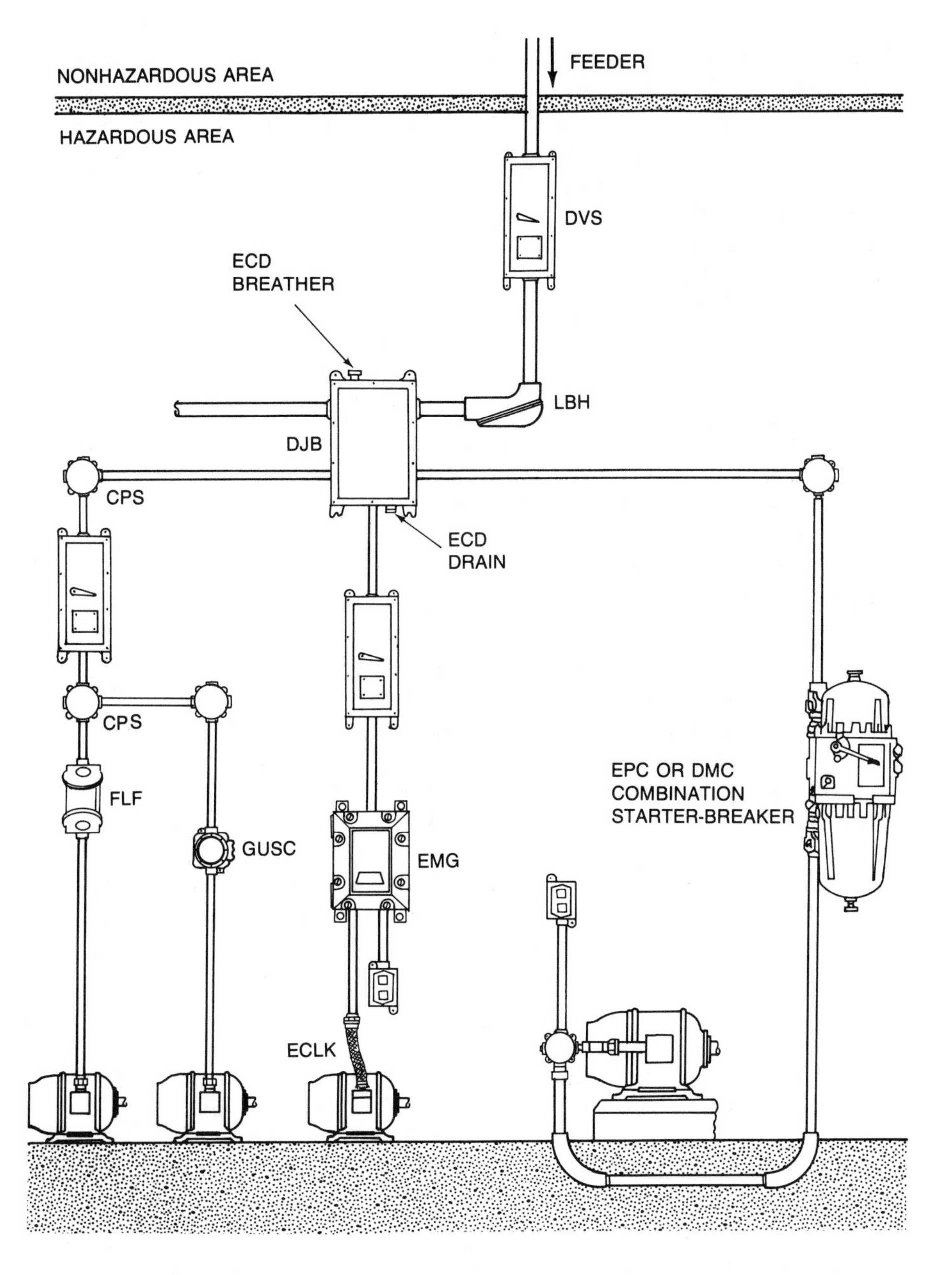

Figure 5-5 Practical representation of explosion-proof fittings. (Courtesy of Crouse-Hinds.)

OUTLET BOXES FOR DAMP AND WET LOCATIONS

In damp or wet locations, boxes and fittings must be placed or equipped to prevent moisture or water from entering and accumulating within the box or fitting. It is recommended that approved boxes of nonconductive material be used with nonmetallic sheathed cable or approved nonmetallic conduit when the cable or conduit is used in locations where there is likely to be occasional moisture present—such as in dairy barns. Boxes installed in wet locations shall be approved for the purpose, according to Section 370-5 of the NE Code.

A wet location is any location subject to saturation with water or other liquids, such as locations exposed to weather, washrooms in garages, and interiors, which might be hosed down. Installations underground or in concrete slabs or masonry in direct contact with the earth must be considered as wet locations. Raintight or watertight equipment may satisfy the requirements for "weatherproof." Boxes with threaded conduit hubs and gasketed covers will normally prevent water from entering the box except for condensation within the box.

A damp location is a location subject to some degree of moisture. Such locations include partially protected outdoor locations—such as under canopies, marquees, and roofed open porches. It also includes interior locations subject to moderate degrees of moisture—such as some basements, some barns, and cold-storage warehouses.

Weatherproof covers for outdoor receptacles must be chosen with care. If the receptacle feeds a permanently connected load (such as a lighting fixture), the entire enclosure must be weatherproof with the plug inserted. If the receptacle is used only with portable tools or other portable equipment, the enclosure must be weatherproof with the cover closed and the cover must be self-closing. Section 410-57 of the NE Code covers installation of receptacles in damp or wet locations.

PULL BOXES

Sheet metal and nonmetallic pull boxes are used as a part of the conduit or tubing raceway system to facilitate the pulling in of the wire or cable, or where it is impractical to install conduit and fittings when change of direction of the raceway run is required. These boxes are also used to provide a junction point for the connection of conductors. Allowance for proper supports for such pull boxes must be made. In some instances, the location and size of pull boxes is designated on the drawings or in the written specifications supplied by the architect or the architect's engineer. However, in most cases, it is the electrician's or contractor's responsibility to size these boxes correctly according to good practices of the NE Code.

Regardless of whether junction or pull boxes are specified for a particular project, long runs of wires should not be made in one pull. Pull boxes, installed at convenient intervals, relieve much of the strain on the wires and make the pull much easier. Since there is no set rule regarding the distance between pull boxes, workers will have to use good judgment regarding where they are required or would be beneficial.

Pull boxes should always be installed in a location that allows workers to work easily and conveniently. In an installation where the conduit run is routed up a corner of a wall and changes direction at the ceiling, a pull box that is installed too high will force the electrician to stand on a ladder when feeding or pulling wires. This situation is just an example; there are several situations to consider when locations for junction boxes are related. They must be securely fastened in place on walls or ceiling or some other adequate support.

To make certain that the raceway system provides a continuous equipment ground, all connections between sections of conduit and between the conduit and termination points (e.g., panels, wire troughs, junction boxes) must be tight. This is insured by the use of two locknuts on every termination point, even though metal bushings are used.

All UL listed metal boxes, except aluminum alloy boxes, are provided with corrosion protection suitable for installation in concrete. Aluminum alloy boxes are not considered acceptable for installation in concrete or cinder fill unless protected by asphalt paint or the equivalent. Boxes designated as "concrete boxes" may have no means of support other than the concrete and often accommodate covers at top and bottom.

6

Wiring Devices

A unit or component of an electrical system intended to carry, but not utilize current, is normally referred to as a *device.*

A *wiring device*, in general terms, is a component of an electrical system to which conductors are attached, which in turn can do any or all of the following: carry, control, or make practical use of electricity possible. Examples of such devices include duplex receptacles, 400-V receptacles, and other types of receptacles, and circuit switches.

Those people involved with the design and installation of wiring devices must be thoroughly familiar with each type and how each is used on actual installations. A natural beginning would be to become familiar with the following definitions:

Cord connector: A cord connector is a portable receptacle, which is provided with means for attachment to a flexible cord and which is not intended for permanent mounting.

Grounded conductor: A grounded conductor is a circuit conductor, which is intentionally connected to an earth ground, and identified as the *white conductor.*

Grounding conductor: A grounding conductor is a conductor, which connects noncurrent-carrying metal parts of equipment to earth ground to provide an intentional path for fault current to ground. It is bare or, when covered, is identified as the *green* or green with yellow stripes conductor.

Lampholder: A lampholder is a device, which is intended to support an electric lamp mechanically and to connect it electrically to a circuit.

Male base (inlet): A male base is a plug, which is intended for flush or surface mounting on an applicance or equipment and which serves to connect utilization equipment to a connector.

Outlet: An outlet is a point on the wiring system at which current is taken to supply utilization equipment.

Plug: A plug is a device with male blades, which, when inserted into a receptacle, establishes connection between the conductors of the attached flexible cord and the conductors connected to the receptacle.

Polarization: Polarization is a means of assuring the mating of plugs and receptacles of the same rating in only the correct position.

Pole: The term *pole* as used in designating plugs and receptacles refers to a terminal to which a circuit conductor (normally current carrying) is connected. In switches, the number of poles indicates the number of conductors being controlled.

Receptacles: A receptacle is a device with female contacts, which is primarily installed at an outlet or on equipment and which is intended to establish electrical connection with an inserted plug.

Slant symbol (/): The "slant" line (/) as used in wiring device ratings indicates that two or more voltage potentials are present simultaneously between different terminals of a wiring device.

Switch: A switch is a device for making, breaking, or changing the connections in an electric circuit.

A. Single-pole (single-pole, single-throw): A switch which makes or breaks the connection of two conductors of a single branch circuit.
B. Double-pole (double-pole, single-throw): A switch which makes or breaks the connection of two conductors of a single branch circuit.
C. Three-way (single-pole, double-throw): A switch which changes the connection of one conductor and which is normally used in pairs to control one utilization equipment from two locations.
D. Four-way (double-pole, double-throw reversing): A form of double-pole switch, which is used in conjunction with two three-way switches to control one utilization equipment from three or more locations.

Terminal: A terminal is a fixed location on a wiring device where a conductor is intended to be connected.

Wire: The term "wire" as used in designating plugs and receptacles indicates the number of either normally current-carrying or equipment-grounding connected conductors.

National Electrical Manufacturer's Association (NEMA) configurations for general-purpose nonlocking plugs and receptacles are shown in Fig. 6-1. NEMA configurations for locking type plugs and receptacles are shown in Fig. 6-2.

Ordering Information

Rating	NEMA Conf.	Color	Catalog Numbers Spec. Grade	Catalog Numbers Hosp. Grade
3 W 15 A 125 V	5-15	Brown	5262DW	85262DW
		Ivory	5262DWI	85262DWI
		Gray	5262DWGRY	85262DWGRY
		Red	—	85262DWR
		White	5262DWW	85262DWW
		Yellow (Corrosion-Resistant)	5262DWCR	—
3 W 15 A 250 V	6-15	Brown	5662DW	—
		Ivory	5662DWI	—
		Yellow (Corrosion-Resistant)	5662DWCR	—
3 W 20 A 125 V	5-20	Brown	5362DW	85362DW
		Ivory	5362DWI	85362DWI
		Gray	5362DWGRY	85362DWGRY
		Red	—	85362DWR
		White	5362DWW	85362DWW
		Yellow (Corrosion-Resistant)	5362DWCR	—
3 W 20 A 250 V	6-20	Brown	5462DW	—
		Ivory	5462DWI	—
		Yellow (Corrosion-Resistant)	5462DWCR	—

Ordering Information for Single Straight-Blade Receptacles

Rating	NEMA Conf.	Color	Spec. Grade	Hosp. Grade
3 W 15 A 125 V	5-15	Brown	5261DW	85261DW
		Ivory	5261DWI	85261DWI
		Gray	5261DWGRY	85261DWGRY
		Red	5261DWR	85261DWR
		Yellow (Corrosion-Resistant)	5261DWCR	85261DWCR
3 W 20 A 125 V	5-20	Brown	5361DW	85361DW
		Ivory	5361DWI	85361DWI
		Gray	5361DWGRY	85361DWGRY
		Red	5361DWR	85361DWR
		Yellow (Corrosion-Resistant)	5361DWCR	—

Figure 6-1 NEMA configurations for general-purpose nonlocking plugs and receptacles.

Ordering Information for TURNEX® Locking Receptacles				
	NEMA		Catalog Numbers	
Rating	Conf.	Color	Spec. Grade	Corr. Resist.
3 W 20 A 125 V	L5-20	Black	4747	
		Yellow		4747CR
3 W 20 A 250 V	L6-20	Black	4748	
		Yellow		4748CR
3 W 20 A 277 V	L7-20	Black	4749	
		Yellow		4749CR
3 W 30 A 125 V	L5-30	Black	4947	
		Yellow		4947CR
3 W 30 A 250 V	L6-30	Black	4948	
		Yellow		4948CR
3 W 30 A 277 V	L7-30	Black	4949	
		Yellow		4949CR
4 W 20 A 125/250 V	L14-20	Black	4774	
		Yellow		4774CR
4 W 20 A 250 V Three-phase	L15-20	Black	4775	
		Yellow		4775CR
4 W 20 A 480 V Three-phase	L16-20	Black	4776	
		Yellow		4776CR
4 W 30 A 125/250 V	L14-30	Black	4974	
		Yellow		4974CR
4 W 30 A 250 V Three-phase	L15-30	Black	4975	
		Yellow		4975CR
4 W 30 A 480 V Three-phase	L16-30	Black	4976	
		Yellow		4976CR

Figure 6-2 NEMA configurations for locking type plugs and receptacles.

SWITCHES

Introduction

Switches used in branch circuit wiring are rated in two general categories: ac only and ac/dc. The original type of electricity furnished years ago was dc, direct current. Also, many industrial applications such as emergency lighting and battery charging circuits also utilize dc current. Because current always flows at full value, the making or breaking of a circuit created a rather strong arc at all times of switch operation, so the contacts had to be gapped widely apart and the operation had to be "quick make, quick break," to avoid excessive pitting or welding of the contact points.

Alternating current (ac) operates between zero and full power 120 times per second (in the United States). The arc from an ac circuit is, because of its oscillating characteristic, self-extinguishing on the "break" of the current, so a slower breaking action is possible with this switch than on one made for use on continuous, full power dc. The pulsating nature of ac can, however, cause a "chattering" or very fast series of "makes" and "breaks" if the circuit is closed (completed) too quickly, so an ac only switch works with a somewhat slower "fast make" than an ac/dc switch.

Because the "make" and "break" actions of an ac/dc switch must be fast, it can be rather noisy, and must use contact metals capable of enduring such comparatively violent action, while still offering acceptable electrical conductivity. An alloy of copper is most often used.

Silver and its oxides are better conductors than copper and its oxides. But silver is much softer than copper and therefore cannot be used in the ac/dc switch. However, silver can be used in the ac only switch because of its gentler "fast make" characteristic, especially when alloyed with cadmium oxide, which is a good conductor and quite hard.

Switch Ratings

Ac only switches, often called ac quiet switches because they can be made very quietly operating (unlike ac/dc switches), are rated at their full current value for lighting (tungsten filament, quartz-metal halide, and fluorescent) and at 80% current value (or full horsepower rating value) for motor (inductive) loads.

Switches that are ac/dc that are used for incandescent lighting should have a *T* rating (tungsten filament rating) that assures the user of longevity of the device through the surge-currents of a "make" action across a tungsten filament. Such capacity is already inherent in the design of ac quiet switches. Switches that are ac/dc are rated at only 50% of their current capacity for motor loads. A 20-A ac/dc switch could be used to control only a 10-A motor, whereas a 20-A ac quiet switch can operate a 16-A motor.

Mercury switches operate by making and breaking contact within a hermetically sealed capsule containing a pool of mercury. Older types were rated like ac/dc switches, but the newer ones are similar in rating to ac quiet switches. They are actually silent in operation.

The obvious limitation of ac quiet and newer mercury switches is that they cannot be used on steady-flow direct current.

Some practical applications of switches are shown in Fig. 6-3.

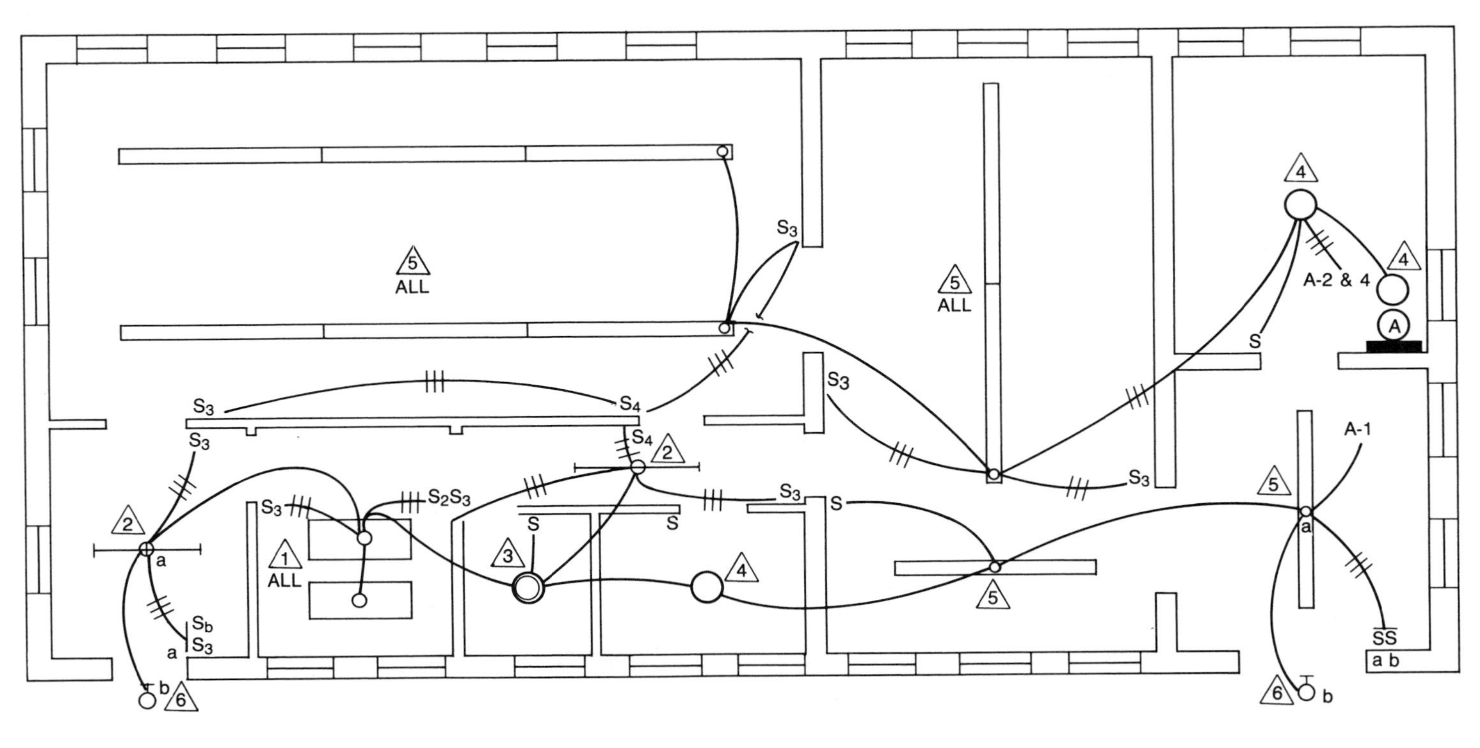

Figure 6-3 Working drawing representing practical applications of switches.

DIMMERS

Dimmers offer the user a variety of lighting effects and a great saving of energy, as well as extended bulb life. Even a simple hi/lo dimmer can save 33% of the electrical energy when used on the low position and can extend bulb life up to 20 times! A full-range dimmer can extend those savings enormously used at still lower settings.

Modern dimmers accomplish their jobs by rectifying the current—changing it from ac to pulsating dc—and thereby reducing its voltage. A more sophisticated kind of control is achieved when the amount of rectification can be varied, as with a *triac* (full-wave silicon controlled rectifier).

RECEPTACLES

Types of Receptacles

Generally, receptacles are made to accommodate either of two kinds of attachments or molded-on plugs. Straight-blade plugs are pushed in and rotated slightly in a clockwise direction to a position in which they cannot be pulled straight out; they must be rotated counterclockwise to the position of insertion to be withdrawn. This locking action is intended to prevent disconnection of the circuit by unintentionally pulling the plug straight out. Obviously, locking devices are used mainly on critical circuits that must not be cut off accidentally.

Grounding

Originally, electrical circuitry was two-pole, two-wire; that is, there was no equipment-grounding means at the receptacle. The only grounding point was at the service entrance, where the neutral or white wire was grounded. The receptacle slot for the white neutral wire, according to NEMA configuration 1-15R, must be longer than the slot for the black or red *phase* or *hot* wire. This enables connection of certain equipment (like TV sets and other appliances) to have their external metal parts or casings grounded through the white neutral wire to the service entrance point. Such items use polarized plugs that have one blade wider than the other. Receptacles for two-pole, two-wire circuits have mounting straps that are not connected to either circuit terminal.

Some years ago it was found that the single grounding point through the neutral conductor was not sufficiently safe; false currents and differences in potential could occur and cause problems such as shocks, overheating, equipment burnout and fires. Therefore, the two-pole, three-wire circuit was developed, wherein the third conductor leads separately to a point intended for equipment grounding. At the receptacle this grounding terminal is most commonly shaped like a *U*; hence the term *U-ground*. The grounding terminals on such receptacles are securely fastened to the mounting strap.

Mounting Straps

There are basically two kinds of mounting straps for receptacles. The *through* type passes between the receptacle body and cover. The *wrap-around* type acts as a cradle for the receptacle, surrounding its top, back, and bottom surfaces. It permits the device to withstand more use and abuse.

Power Contact Design

The contact must grip the plug blade—not merely lie against it—for maximum electrical conductivity and mechanical strength. Some receptacles have double-wipe contacts that grip both sides of each blade, including the U-ground contacts; this type normally combines maximum strength and electrical conductivity.

GROUND-FAULT CIRCUIT INTERRUPTERS

While circuit breakers and fuses protect equipment and structures against high-current overloads and short circuits, they do not protect personnel against electrocution. A person can be killed by as small a current as 200 ma, and sometimes smaller! The most common type of current responsible for electrocutions is the *ground fault*, a leakage of current to ground, often through the body of a person in contact in some way with ground, thus providing the ground path.

A ground-fault circuit interrupter (GFCI) will detect such leakage and will open the circuit promptly to avoid further shock or injury to the person.

Such a device can be placed at the distribution center, usually as part of a dual-function circuit breaker. UL requires that a GFCI will trip from either this kind of ground fault or from an overcurrent in the circuit greater than its rating (e.g., 15 or 20 A). Troubleshooting this trip is not often easy, since either a ground fault or an overcurrent could have caused the trip, and since the entire branch is shut off by the circuit breakers.

Today there are receptacles available with built-in GFCI protection. They can be installed to offer GFCI protection only through their own outlets, or they can be used as feed-through devices protecting a part of or the entire branch circuit.

7

Electrical Conductors

INTRODUCTION

A variety of materials is used to transmit electrical energy, but copper—due to its excellent cost-to-conductivity ratio—still remains the basic and most ideal conductor. Electrolytic copper, the type used in most electrical conductors, can have three general characteristics:

1. Method of stranding
2. Degree of hardness (temper)
3. Bare, tinned, or coated

Method of Stranding

Stranding refers to the relative flexibility of the conductor and may consist of only one strand or many thousands, depending on the rigidity or flexibility required for a specific need. For example, a small-gauge wire that is to be used in a fixed installation is normally solid (one strand), whereas a wire that will be constantly flexed requires a high degree of flexibility and would contain many strands.

1. Solid is the least flexible form of a conductor and is merely one strand of copper.
2. Stranded refers to more than one strand in a given conductor and may vary from 3 to 19 depending on size.
3. Flexible simply indicates that there are a greater number of strands than are found in normal stranded construction.

Degree of Hardness (Temper)

Temper refers to the relative hardness of the conductor and is noted as soft drawn-annealed (SD), medium hard drawn (MHD), and hard drawn (HD). Again, the specific need of an installation will determine the required temper. Where greater tensile strength is indicated, MHD would be specified over SD, and so on.

Bare, Tinned, or Coated

Untinned copper is plain bare copper that is available in either solid, stranded, or flexible and in the various tempers just described. In this form it is often referred to as *red* copper.

Bare copper is also available with a coating of tin, silver, or nickel to facilitate soldering, to impede corrosion, and to prevent adhesion of the copper conductor to rubber or other types of conductor insulation. The various coatings will also affect the electrical characteristics of copper.

CONDUCTOR SIZE

The American Wire Gauge (AWG) is used in the United States to identify the size of wire and cable up to and including No. 4/0 (0000), which is commonly pronounced in the electrical trade as "four-aught" or "four-naught." These numbers run in reverse order as to size; that is, No. 14 AWG is smaller than No. 12 AWG and so on up to size No. 1 AWG. To this size (No. 1 AWG), the larger the gauge number, the smaller the size of the conductor. However, the next larger size after No. 1 AWG is No. 1/0 AWG, then 2/0 AWG, 3/0 AWG, and 4/0 AWG. At this point, the AWG designations end and the larger sizes of conductors are identified by circular mils (cmil). From this point, the larger the size of wire, the larger the number of circular mils. For example, 300,000 cmil is larger than 250,000 cmil. In writing these sizes in circular mils, the "thousand" decimal is replaced by the letter M, and instead of writing, say, 500,000 cmil, it is usually written MCM.

TYPES OF CONDUCTORS

The following types of insulated conductors are for normal use in electrical systems of 600 V or less. The maximum continuous ampacities for these conductors are specified in Tables 310-6 through 310-10 of the NE Code.

Conductor Type	Maximum Operating Temperature (°C)	Application
A	200 (392°F)	Dry locations only
AA	200 (392°F)	Dry locations only
AI	125 (257°F)	Dry locations only
AIA	125 (257°F)	Dry locations only
AVA	110 (185°F)	Dry locations only
AVB	90 (194°F)	Dry locations only
AVL	110 (185°F)	Dry locations only
FEP	90 (194°F)	Dry locations
FEPW	75 (167°F)	Wet locations
MI	85 (185°F)	Dry and wet locations
MTW	60 (140°F)	Machine tool wiring in wet areas
PFA	90 (194°F)	Dry locations
PFAH	250 (482°F)	Dry locations only
Paper	85 (185°F)	For underground service
RH	75 (167°F)	For use in dry locations
RHH	90 (194°F)	For use in dry locations
RHW	75 (167°F)	Dry and wet locations
RUH	75 (167°F)	Dry locations
RUW	60 (140°F)	Dry and wet locations
SA	90 (194°F)	Dry locations
SIS	90 (194°F)	Switchboard wiring only
T	60 (140°F)	Dry locations
TA	90 (194°F)	Switchboard wiring only
TBS	90 (194°F)	Switchboard wiring only
TEE	250 (482°F)	Dry locations only
THNN	90 (194°F)	Dry locations
THW	75 (167°F)	Dry and wet locations
THWN	75 (167°F)	Dry and wet locations
TW	60 (140°F)	Dry and wet locations
UF	60 (140°F)	See Article 339 of NE Code
USE	75 (167°F)	See Article 338 of NE Code
V	85 (185°F)	Dry and wet locations
XHHN	75 (167°F)	Dry and wet locations
Z	90 (194°F)	Dry locations
ZW	75 (167°F)	Wet locations

The Tables 7-1 through 7-11 give characteristics of various sizes and types of electrical conductors. In using these tables, however, the following should be noted:

1. *Aluminum conductors:* For aluminum conductors, the allowable current-carrying capacities shall be taken as 84% of those given in the table for the respective sizes of copper conductor with the same kind of insulation.
2. *Bare conductors:* If bare conductors are used with insulated con-

ductors, their allowable current-carrying capacity shall be limited to that permitted for the insulated conductor with which they are used.

3. *Application of tables:* For most types of wiring, the allowable current-carrying capacities of Table 7-1 may be used, unless otherwise provided in the NE Code.
4. *More than three conductors in a raceway:* Table 7-1 gives the allowable current-carrying capacity for not more than three conductors in a raceway or cable. If the number of conductors in a raceway or cable is from 4 to 6, the allowable current-carrying capacity of each conductor shall be reduced to 80% of the values in Table 7-1. If the number of conductors in a raceway or cable is from 7 to 9, the allowable current-carrying capacity of each conductor shall be reduced to 70% of the values in Table 7-1.
5. *Neutral conductor:* A neutral conductor that carries only the unbalanced current from other conductors, as in the case of normally balanced circuits of three or more conductors, shall not be counted in determining current-carrying capacities as provided for in the preceding paragraph.
6. *Ultimate insulation temperature:* In no case shall conductors be associated together in such a way with respect to the kind of circuit, the wiring method employed, or the number of conductors, that the limiting temperature of the conductors will be exceeded.
7. *Use of conductors with higher operating temperatures:* If the room temperature is within 10°C of the maximum allowable operating temperature of the insulation, it is desirable to use an insulation with a higher maximum allowable operating temperature; although insulation can be used in a room temperature approaching its maximum allowable operating temperature limit if the current is reduced in accordance with the table of correction factors for different room temperatures.
8. *Voltage drop:* The allowable current-carrying capacities in Tables 7-1 and 7-2 are based on temperature alone and do not take voltage drop into consideration.
9. *Overcurrent protection:* If the standard ratings and settings of overcurrent devices do not correspond with the ratings and settings allowed for conductors, the next higher standard rating and setting may be used, but it cannot exceed 150% of the allowable capacity of the conductor.
10. *Deterioration of insulation:* It should be noted that even the best grades of rubber insulation will deteriorate in time and will eventually need to be replaced.

TABLE 7-1 ALLOWABLE CURRENT-CARRYING CAPACITIES OF INSULATED COPPER CONDUCTORS (A)

Size AWG (MCM)	Conductor Type					
	Rubber [Types R, RW, RU, RUW (14-2)] Type RH-RW (Note) Thermo-plastic Types T, TW	Rubber Type RH [Type RUH (14-2)] Type RH-RW (Note) Type RHW	Paper Thermo-plastic Asbestos, Type TA Var-Cam Type V Asbestos Var-Cam Type AVB MI Cable RHH*	Asbestos Var-Cam Types AVA, AVL	Impreg-nated Asbestos [Type AI (14-8)] Type AIA	Asbestos [Type A (14-8)] Type AA
14	15	15	25	30	30	30
12	20	20	30	35	40	40
10	30	30	40	45	50	55
8	40	45	50	60	65	70
6	55	65	70	80	85	95
4	70	85	90	105	115	120
3	80	100	105	120	130	145
2	95	115	120	135	145	165
1	110	130	140	160	170	190
0	125	150	155	190	200	225
00	145	175	185	215	230	250
000	165	200	210	245	265	285
0000	195	230	235	275	310	340
250	215	255	270	315	335	—
300	240	285	300	345	380	—
350	260	310	325	390	420	—
400	280	335	360	420	450	—

500	320	380	405	470	500	—
600	355	420	455	525	545	—
700	385	460	490	560	600	—
750	400	475	500	580	620	—
800	410	490	515	600	640	—
900	435	520	555	—	—	—
1000	455	545	585	680	730	—
1250	495	590	645	—	—	—
1500	520	625	700	785	—	—
1750	545	650	735	—	—	—
2000	560	665	775	840	—	—

Correction Factors [Room Temperatures Over 30°C (86°F)]

C	F						
40	(104°F)	.82	.88	.90	.94	.95	—
45	(113°F)	.71	.82	.85	.90	.92	—
50	(122°F)	.58	.75	.80	.87	.89	—
55	(131°F)	.41	.67	.74	.83	.86	—
60	(140°F)	—	.58	.67	.79	.83	.91
70	(158°F)	—	.35	.52	.71	.76	.87
75	(167°F)	—	—	.43	.66	.72	.86
80	(176°F)	—	—	.30	.61	.69	.84
90	(194°F)	—	—	—	.50	.61	.80
100	(212°F)	—	—	—	—	.51	.77
120	(248°F)	—	—	—	—	—	.69
140	(284°F)	—	—	—	—	—	.59

*The current-carrying capacities for Type RHH conductors for sizes AWG 14, 12, and 10 shall be the same as designated for Type RH conductors in this table.

Note: Not more than three conductors in raceway or cable or direct burial [based on room temperature of 30°C (86°F)].

TABLE 7-2 ALLOWABLE CURRENT-CARRYING CAPACITIES OF INSULATED ALUMINUM CONDUCTORS (A)

	Conductor Type					
Size AWG (MCM)	Rubber [Types R, RW, RU, RUW (12-2)] Type RH-RW (Note) Thermo-plastic Types T, TW	Rubber Type RH [Type RUH (14-2)] Type RH-RW (Note) Type RHW	Paper Thermo-plastic Asbestos, Type TA Var-Cam, Type V Asbestos Var-Cam Type AVB MI Cable RHH*	Asbestos Var-Cam Types AVA, AVL	Impreg-nated Asbestos [Type AI (14-8)] Type AIA	Asbestos [Type A (14-8)] Type AA
12	15	15	25	25	30	30
10	25	25	30	35	40	45
8	30	40	40	45	50	55
6	40	50	55	60	65	75
4	55	65	70	80	90	95
3	65	75	80	95	100	115
+2	75	90	95	105	115	130
+1	85	100	110	125	135	150
+0	100	120	125	150	160	180
+00	115	135	145	170	180	200
+000	130	155	165	195	210	225
+0000	155	180	185	215	245	270
250	170	205	215	250	270	—
300	190	230	240	275	305	—
350	210	250	260	310	335	—
400	225	270	290	335	360	—
500	260	310	330	380	405	—
600	285	340	370	425	440	—

700	310	375	395	455	485	—
750	320	385	405	470	500	—
800	330	395	415	485	520	—
900	355	425	455	—	—	—
1000	375	445	480	560	600	—
1250	405	485	530	—	—	—
1500	435	520	580	650	—	—
1750	455	545	615	—	—	—
2000	470	560	650	705	—	—
Correction Factors [Room Temperatures Over 30°C (86°F)]						
C F						
40 (104°F)	.82	.88	.90	.94	.95	—
45 (113°F)	.71	.82	.85	.90	.92	—
50 (122°F)	.58	.75	.80	.87	.89	—
55 (131°F)	.41	.67	.74	.83	.86	—
60 (140°F)	—	.58	.67	.79	.83	.91
70 (158°F)	—	.35	.52	.71	.76	.87
75 (167°F)	—	—	.43	.66	.72	.86
80 (176°F)	—	—	.30	.61	.69	.84
90 (194°F)	—	—	—	.50	.61	.80
100 (212°F)	—	—	—	—	.51	.77
120 (248°F)	—	—	—	—	—	.69
140 (284°F)	—	—	—	—	—	.59

Note: Not more than three conductors in raceway or cable or direct burial [based on room temperature of 30°C (86°F)].

*The current-carrying capacities for Type RHH conductors for sizes AWG 12, 10, and 8 shall be the same as designated for Type RH conductors in this table.

+For three-wire, single-phase service and sub-service circuits, the allowable current-carrying capacity of RH, RH-RW, RHH, and RHW aluminum conductors shall be for sizes #2-100 A, #1-110 A, #1/0-125 A, #2/0-150 A, #3/0-170 A, and #4/0-200 A.

TABLE 7-3 ALLOWABLE CURRENT-CARRYING CAPACITIES OF INSULATED COPPER CONDUCTORS (A)

Size AWG (MCM)	Conductor Type						
	Rubber [Types R, RW, RU, RUW (14-2)] Type RH-RW (Note) Thermo-plastic Types T, TW	Rubber Type RH [Type RUH (14-2)] Type RH-RW (Note) Type RHW	Paper Thermo-plastic Asbestos, Type TA Var-Cam, Type V Asbestos Var-Cam Type AVB MI Cable RHH*	Asbestos Var-Cam Types AVA, AVL	Impreg-nated Asbestos [Type AI (14-8)] Type AIA	Asbestos [Type A (14-8)] Type AA	Slow-burning Type SB Weather-proof Type WP
14	20	20	30	40	40	45	30
12	25	25	40	50	50	55	40
10	40	40	55	65	70	75	55
8	55	65	70	85	90	100	70
6	80	95	100	120	125	135	100
4	105	125	135	160	170	180	130
3	120	145	155	180	195	210	150
2	140	170	180	210	225	240	175
1	165	195	210	245	265	280	205
0	195	230	245	285	305	325	235
00	225	265	285	330	355	370	275
000	260	310	330	385	410	430	320
0000	300	360	385	445	475	510	370
250	340	405	425	495	530	—	410
300	375	445	480	555	590	—	460
350	420	505	530	610	655	—	510
400	455	545	575	665	710	—	555

500		515	620	660	765	815	—	630
600		575	690	740	855	910	—	710
700		630	755	815	940	1005	—	780
750		655	785	845	980	1045	—	810
800		680	815	880	1020	1085	—	845
900		730	870	940	—	—	—	905
1000		780	935	1000	1165	1240	—	965
1250		890	1065	1130	—	—	—	—
1500		980	1175	1260	1450	—	—	1215
1750		1070	1280	1370	—	—	—	—
2000		1155	1385	1470	1715	—	—	1405
Correction Factors [Room Temperatures Over 30°C (86°F)]								
C	F							
40	(104°F)	.82	.88	.90	.94	.95	—	—
45	(113°F)	.71	.82	.85	.90	.92	—	—
50	(122°F)	.58	.75	.80	.87	.89	—	—
55	(131°F)	.41	.67	.74	.83	.86	—	—
60	(140°F)	—	.58	.67	.79	.83	.91	—
70	(158°F)	—	.35	.52	.71	.76	.87	—
75	(167°F)	—	—	.43	.66	.72	.86	—
80	(176°F)	—	—	.30	.61	.69	.84	—
90	(194°F)	—	—	—	.50	.61	.80	—
100	(212°F)	—	—	—	—	.51	.77	—
120	(248°F)	—	—	—	—	—	.69	—
140	(284°F)	—	—	—	—	—	.59	—

*The current-carrying capacities for Type RHH conductors for sizes AWG 14, 12, and 10 shall be the same as designated for Type RH conductors in this table.

Note: Single conductor in free air [based on room temperature of 30°C (86°F)].

TABLE 7-4 ALLOWABLE CURRENT-CARRYING CAPACITIES OF INSULATED ALUMINUM CONDUCTORS (A)

	Conductor Type						
Size AWG (MCM)	Rubber [Types R, RW, RU, RUW (12-2)] Type RH-RW (Note) Thermo-plastic Types T, TW	Rubber Type RH [Type RUH (14-2)] Type RH-RW (Note) Type RHW	Paper Thermo-plastic Asbestos, Type TA Var-Cam, Type V Asbestos Var-Cam Type AVB MI Cable RHH*	Asbestos Var-Cam Types AVA, AVL	Impreg-nated Asbestos [Type AI (14-8)] Type AIA	Asbestos [Type A (14-8)] Type AA	Slow-burning Type SB Weather-proof Type WP
12	20	20	30	40	40	45	30
10	30	30	45	50	55	60	45
8	45	55	55	65	70	80	55
6	60	75	80	95	100	105	80
4	80	100	105	125	135	140	100
3	95	115	120	140	150	165	115
2	110	135	140	165	175	185	135
1	130	155	165	190	205	220	160
0	150	180	190	220	240	255	185
00	175	210	220	255	275	290	215
000	200	240	255	300	320	335	250
0000	230	280	300	345	370	400	290
250	265	315	330	385	415	—	320
300	290	350	375	435	460	—	360
350	330	395	415	475	510	—	400
400	355	425	450	520	555	—	435

500	405	485	515	595	635	—	490
600	455	545	585	675	720	—	560
700	500	595	645	745	795	—	615
750	515	620	670	775	825	—	640
800	535	645	695	805	855	—	670
900	580	700	750	—	—	—	725
1000	625	750	800	930	990	—	770
1250	710	855	905	—	—	—	—
1500	795	950	1020	1175	—	—	985
1750	875	1050	1125	—	—	—	—
2000	960	1150	1220	1425	—	—	1165
Correction Factors [Room Temperatures Over 30°C (86°F)]							
C F							
40 (104°F)	.82	.88	.90	.94	.95	—	—
45 (113°F)	.71	.82	.85	.90	.92	—	—
50 (122°F)	.58	.75	.80	.87	.89	—	—
55 (131°F)	.41	.67	.74	.83	.86	—	—
60 (140°F)	—	.58	.67	.79	.83	.91	—
70 (158°F)	—	.35	.52	.71	.76	.87	—
75 (167°F)	—	—	.43	.66	.72	.86	—
80 (176°F)	—	—	.30	.61	.69	.84	—
90 (194°F)	—	—	—	.50	.61	.80	—
100 (212°F)	—	—	—	—	.51	.77	—
120 (248°F)	—	—	—	—	—	.69	—
140 (284°F)	—	—	—	—	—	.59	—

*The current-carrying capacities for Type RHH conductors for sizes AWG 12, 10, and 8 shall be the same as designated for Type RH conductors in this table.

+For three-wire, single-phase service and sub-service circuits, the allowable current-carrying capacity of RH, RH-RW, RHH, and RHW aluminum conductors shall be for sizes #2-100 amp, #1-110 amp, #1/0-125 amp, #2/0-150 amp, #3/0-170 amp, and #4/0-200 amp.

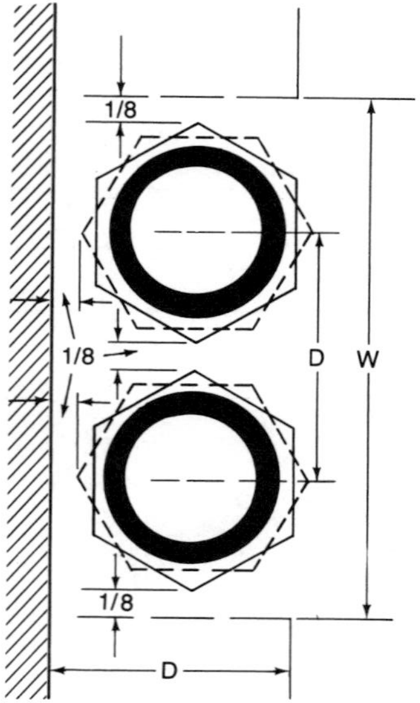

TABLE 7-5 SPACE REQUIREMENTS FOR CONDUITS (IN.-DIAMETER)

Conduit Size (in.)	Space Requirement (in.)							
	1/2		3/4		1		1 1/4	
	D	W	D	W	D	W	D	W
1/2	1 1/4	2 5/8	1 3/8	2 7/8	1 1/2	3 1/8	1 11/16	3 1/2
3/4	1 3/8	2 7/8	1 1/2	3 1/8	1 5/8	3 3/8	1 13/16	3 3/4
1	1 1/2	3 1/8	1 5/8	3 3/8	1 3/4	3 5/8	1 15/16	4
1 1/4	1 11/16	3 1/2	1 13/16	3 3/4	1 15/16	4	2 1/8	4 3/8
1 1/2	1 13/16	3 3/4	1 15/16	4	2 1/16	4 1/4	2 1/4	4 5/8
2	2 1/8	4 3/8	2 1/4	4 5/8	2 3/8	4 7/8	2 3/8	4 7/8
2 1/2	2 7/16	5	2 9/16	5 1/4	2 11/16	5 1/2	2 7/8	5 7/8
3	2 3/4	5 5/8	3	6 1/8	3	6 1/8	3 1/4	6 5/8
3 1/2	3 1/8	6 3/8	3 1/4	6 5/8	3 3/8	6 7/8	3 9/16	7 1/4

Conduit Size (in.)	Space Requirement (in.)									
	1 1/2		2		2 1/2		3		3 1/2	
	D	W	D	W	D	W	D	W	D	W
1/2	1 13/16	3 3/4	2 1/8	4 3/8	2 7/16	5	2 3/4	5 5/8	3 1/8	6 3/8
3/4	1 15/16	4	2 1/4	4 5/8	2 9/16	5 1/4	3	6 1/8	3 1/4	6 5/8
1	2 1/16	4 3/8	2 3/8	4 7/8	2 11/16	5 1/2	3	6 1/8	3 3/8	6 7/8
1 1/4	2 1/4	4 5/8	2 9/16	5 1/4	2 7/8	5 7/8	3 1/4	6 5/8	3 9/16	7 1/4
1 1/2	2 3/8	4 7/8	2 11/16	5 1/2	3	6 1/8	3 3/8	6 7/8	3 11/16	7 1/2
2	2 11/16	5 1/2	3	6 1/8	3 5/16	6 3/4	3 5/8	7 3/8	4	8 1/8
2 1/2	3	6 1/8	3 5/16	6 3/4	3 5/8	7 3/8	3 15/16	8	4 5/16	8 3/4
3	3 3/8	6 7/8	3 5/8	7 3/8	3 15/16	8	4 7/16	9	4 3/4	9 5/8
3 1/2	3 11/16	7 1/2	4	8 1/8	4 5/16	8 3/4	4 3/4	9 5/8	5	10 1/8

TABLE 7-6 NUMBER OF CONDUCTORS ALLOWED IN CONDUIT OR TUBING

Size AWG (MCM)	Number of Conductors in One Conduit or Tubing								
	1	2	3	4	5	6	7	8	9
18	1/2	1/2	1/2	1/2	1/2	1/2	1/2	3/4	3/4
16	1/2	1/2	1/2	1/2	1/2	1/2	3/4	3/4	3/4
14	1/2	1/2	1/2	1/2	3/4	3/4	1	1	1
12	1/2	1/2	1/2	3/4	3/4	1	1	1	1 1/4
10	1/2	3/4	3/4	3/4	1	1	1	1 1/4	1 1/4
8	1/2	3/4	3/4	1	1 1/4	1 1/4	1 1/4	1 1/2	1 1/2
6	1/2	1	1	1 1/4	1 1/2	1 1/2	2	2	2
4	1/2	1 1/4	*1 1/4	1 1/2	1 1/2	2	2	2	2 1/2
3	3/4	1 1/4	1 1/4	1 1/2	2	2	2	2 1/2	2 1/2
2	3/4	1 1/4	1 1/4	2	2	2	2 1/2	2 1/2	2 1/2
1	3/4	1 1/2	1 1/2	2	2 1/2	2 1/2	2 1/2	3	3
0	1	1 1/2	2	2	2 1/2	2 1/2	3	3	3
00	1	2	2	2 1/2	2 1/2	3	3	3	3 1/2
000	1	2	2	2 1/2	3	3	3	3 1/2	3 1/2
0000	1 1/4	2	2 1/2	3	3	3	3 1/2	3 1/2	4

250	1 1/4	2 1/2	2 1/2	3	3	3 1/2	4	4	5
300	1 1/4	2 1/2	2 1/2	3	3 1/2	4	4	5	5
350	1 1/4	3	3	3 1/2	3 1/2	4	5	5	5
400	1 1/2	3	3	3 1/2	4	4	5	5	5
500	1 1/2	3	3	3 1/2	4	5	5	5	6
600	2	3 1/2	3 1/2	4	5	5	6	6	6
700	2	3 1/2	3 1/2	5	5	5	6	6	—
750	2	3 1/2	3 1/2	5	5	6	6	6	—
800	2	3 1/2	4	5	5	6	6	—	—
900	2	4	4	5	6	6	6	—	—
1000	2	4	4	5	6	6	—	—	—
1250	2 1/2	5	5	6	6	—	—	—	—
1500	3	5	5	6	—	—	—	—	—
1750	3	5	6	6	—	—	—	—	—
2000	3	6	6	—	—	—	—	—	—

Note: Table includes information on rubber-covered types RF-2, RFH-2, R, RH, RHH, RHW, RW, RH-RW, RU, RUH, and RUW; and thermoplastic types TF, T, and TW conductors.

*Where a service run of conduit or electrical metallic tubing does not exceed 50′ in length and does not contain more than the equivalent of two quarter bends from end to end, two No. 4 insulated and one No. 4 bare conductors may be installed in 1″ conduit or tubing.

TABLE 7-7 NUMBER OF CONDUCTORS IN CONDUIT OR TUBING

Size AWG (MCM)	Number of Conductors in One Conduit or Tubing											
	Single-Conductor Cable				Two-Conductor Cable				Three-Conductor Cable			
	1	2	3	4	1	2	3	4	1	2	3	4
14	1/2	3/4	3/4	1	3/4	1	1	1 1/4	3/4	1 1/4	1 1/2	1 1/2
12	1/2	3/4	3/4	1	3/4	1	1 1/4	1 1/4	1	1 1/4	1 1/2	2
10	1/2	3/4	1	1	3/4	1 1/4	1 1/4	1 1/2	1	1 1/2	2	2
8	1/2	1	1 1/4	1 1/2	1	1 1/4	1 1/2	2	1	2	2	2 1/2
6	3/4	1 1/4	1 1/2	1 1/2	1 1/4	1 1/2	2	2 1/2	1 1/4	2 1/2	3	3
4	3/4	1 1/4	1 1/2	1 1/2	1 1/4	2	2 1/2	2 1/2	1 1/2	3	3	3 1/2
3	3/4	1 1/4	1 1/2	2	1 1/4	2	2 1/2	3	1 1/2	3	3	3 1/2
2	1	1 1/4	1 1/2	2	1 1/4	2	2 1/2	3	1 1/2	3	3 1/2	4
1	1	1 1/2	2	2	1 1/2	2 1/2	3	3 1/2	2	3 1/2	4	5
0	1	2	2	2 1/2	2	2 1/2	3	3 1/2	2	4	5	5
00	1	2	2	2 1/2	2	3	3 1/2	4	2 1/2	4	5	5
000	1 1/4	2	2 1/2	2 1/2	2	3	3 1/2	4	2 1/2	5	5	6

0000	1 1/4	2 1/2	2 1/2	3	2 1/2	3	3 1/2	5	3	5	6	6
250	1 1/4	2 1/2	3	3	—	—	—	—	3	6	6	—
300	1 1/2	3	3	3 1/2	—	—	—	—	3 1/2	6	6	—
350	1 1/2	3	3	3 1/2	—	—	—	—	3 1/2	6	6	—
400	1 1/2	3	3	3 1/2	—	—	—	—	3 1/2	6	6	—
500	1 1/2	3	3 1/2	4	—	—	—	—	4	6	—	—
600	2	3 1/2	4	5	—	—	—	—	—	—	—	—
700	2	4	4	5	—	—	—	—	—	—	—	—
750	2	4	4	5	—	—	—	—	—	—	—	—
800	2	4	5	5	—	—	—	—	—	—	—	—
900	2 1/2	4	5	5	—	—	—	—	—	—	—	—
1000	2 1/2	5	5	6	—	—	—	—	—	—	—	—
1250	3	5	5	6	—	—	—	—	—	—	—	—
1500	3	5	6	6	—	—	—	—	—	—	—	—
1750	3	6	6	—	—	—	—	—	—	—	—	—
2000	3 1/2	6	6	—	—	—	—	—	—	—	—	—

Note: Table includes information on conductors that are lead-covered types RL and RHL; 600 V.

The above sizes apply to straight runs or runs with nominal offsets equivalent to not more than two quarter bends.

TABLE 7-8 DIMENSIONS OF RUBBER-COVERED AND THERMOPLASTIC-COVERED CONDUCTORS

Size AWG (MCM)	Types RF-2, RFH-2, R, RH, RHH, RHW, RH-RW, RW		Types TF, T, TW, RU*, RUH*, RUW	
	Diam (in.)	Area (in.^{2})	Diam (in.)	Area (in.^{2})
18	.146	.0167	.106	.0088
16	.158	.0196	.118	.0109
14	3/64 in. .171	.0230	.131	.0135
14	3/64 in. .204+	.0327+	—	—
12	2/64 in. .188	.0278	.148	.0172
12	3/64 in. .221+	.0384+	—	—
10	.242	.0460	.168	.0224
8	.311	.0760	.228	.0408
6	.397	.1238	.323	.0819
4	.452	.1605	.372	.1087
3	.481	.1817	.401	.1263
2	.513	.2067	.433	.1473
1	.588	.2715	.508	.2027
0	.629	.3107	.549	.2367
00	.675	.3578	.595	.2781
000	.727	.4151	.647	.3288
0000	.785	.4840	.705	.3904
250	.868	.5917	.788	.4877
300	.933	.6837	.843	.5581
350	.985	.7620	.895	.6291
400	1.032	.8365	.942	.6969
500	1.119	.9834	1.029	.8316
600	1.233	1.1940	1.143	1.0261
700	1.304	1.3355	1.214	1.1575
750	1.339	1.4082	1.249	1.2252
800	1.372	1.4784	1.282	1.2908
900	1.435	1.6173	1.345	1.4208
1000	1.494	1.7531	1.404	1.5482
1250	1.676	2.2062	1.577	1.9532
1500	1.801	2.5475	1.702	2.2748
1750	1.916	2.8895	1.817	2.5930
2000	2.021	3.2079	1.922	2.9013

*No. 14 to No. 2.

No. 18 to No. 8, solid; No. 6 and larger, stranded.

+The dimensions of Types RW wire and RHH wire. Also, these dimensions to be used for new work in computing size of conduit or tubing for combinations of wires not shown in Table 4.

TABLE 7-9 DIMENSIONS OF LEAD-COVERED CONDUCTORS

Size AWG (MCM)	Single-Conductor Cable		Two-Conductor Cable		Three-Conductor Cable	
	Diam. (in.)	Area (in.^{2})	Diam. (in.)	Area (in.^{2})	Diam. (in.)	Area (in.^{2})
14	.28	.062	.28 X .47	.115	.59	.273
12	.29	.066	.31 X .54	.146	.62	.301
10	.35	.096	.35 X .59	.180	.68	.363
8	.41	.132	.41 X .71	.255	.82	.528
6	.49	.188	.49 X .86	.369	.97	.738
4	.55	.237	.54 X .96	.457	1.08	.916
2	.60	.283	.61 X 1.08	.578	1.21	1.146
1	.67	.352	.70 X 1.23	.756	1.38	1.49
0	.71	.396	.74 X 1.32	.859	1.47	1.70
00	.76	.454	.79 X 1.41	.980	1.57	1.94
000	.81	.515	.84 X 1.52	1.123	1.69	2.24
0000	.87	.593	.90 X 1.64	1.302	1.85	2.68
250	.98	.754	—	—	2.02	3.20
300	1.04	.85	—	—	2.15	3.62
350	1.10	.95	—	—	2.26	4.02
400	1.14	1.02	—	—	2.40	4.52
500	1.23	1.18	—	—	2.59	5.28

Notes: No. 14 to No. 8, solid conductors; No. 6 and larger, stranded conductors.
Data for 2/64″ insulation not yet compiled.
Table includes information on Type RL, RHL, and RUL conductors.

TABLE 7-10 APPLICATION OF CONDUCTORS

Trade Name	Type Letter	Max. Oper. Temp. (°C)	Insulation Type	Outer Cover	Special Provisions and Use
Rubber-covered fixture wire solid or seven-strand	RF-1	60 (140°F)	Code rubber	Nonmetallic covering	Fixture wiring limited to 300 V
	RF-2	60 (140°F)	Code rubber	Nonmetallic covering	Fixture wiring and as permitted
Rubber-covered fixture wire flexible stranding	FF-1	60 (140°F)	Code rubber	Nonmetallic covering	Fixture wiring limited to 300 V in Section 725-14
	FF-2	60 (140°F)	Code rubber	Nonmetallic covering	Fixture wiring and as permitted in Section 725-14
Heat-resistant silicone solid or seven-strand	SF-1	200 (392°F)	Silicone rubber	Lacquered glass	High temperature fixture wiring limited to 300 V
	SF-2	200 (392°F)	Silicone rubber	Lacquered glass	High temperature fixture wiring and as permitted in Section 725-14
Heat-resistant rubber-covered fixture wire—solid or seven-strand	RFH-1	75 (167°F)	Heat resistant rubber	Nonmetallic covering	Fixture wiring limited to 300 V
	RFH-2	75 (167°F)	Heat resistant rubber	Nonmetallic covering	Fixture wiring and as permitted in Section 725-14
Heat resistant rubber-covered fixture wire flexible stranding	FFH-1	75 (167°F)	Heat resistant rubber	Nonmetallic covering	Fixture wiring limited to 300 V
	FFH-2	75 (167°F)	Heat resistant rubber	Nonmetallic covering	Fixture wiring and as permitted in Section 725-14
Thermoplastic-covered fixture wire—solid or stranded	TF	60 (140°F)	Thermoplastic	None	Fixture wiring and as permitted in Section 725-14

Thermoplastic-covered fixture wire—flexible stranding	TFF	60 (140°F)	Thermoplastic	None	Fixture wiring and as permitted in Section 725-14
Cotton-covered heat-resistant fixture wire	CF	90 (194°F)	Impregnated cotton	None	Fixture wiring limited to 300 V
Asbestos-covered heat-resistant fixture wire	AF	150 (302°F)	Impregnated asbestos	None	Fixture wiring limited to 300 V. Indoor dry location.
Heat-resistant rubber	RH	75 (167°F)	Heat-resistant rubber	Moisture-resistant Flame-retardant Nonmetallic covering	Dry locations
Heat-resistant rubber	RHH	90 (194°F)	Heat-resistant rubber	Moisture-resistant Flame-retardant Fibrous covering	Dry locations
Moisture-resistant rubber	RW	60 (140°F)	Moisture-resistant rubber	Moisture-resistant Flame-retardant Nonmetallic covering	Dry and wet locations
Heat and moisture-resistant rubber	RHW	75 (167°F)	Heat and moisture-resistant rubber	Moisture-resistant Flame-retardant Nonmetallic covering	Dry locations
Heat-resistant latex rubber	RUH	75 (167°F)	90% unmilled grainless rubber	Moisture-resistant Flame-retardant Nonmetallic covering	Dry locations
Moisture-resistant latex rubber	RUW	60 (140°F)	90% unmilled grainless rubber	Moisture-resistant Flame-retardant Nonmetallic covering	Dry and wet locations
Thermoplastic	T	60 (140°F)	Flame-retardant thermoplastic compound	None	Dry locations

TABLE 7-11 (cont.)

Trade Name	Type Letter	Max. Oper. Temp. (°C)	Insulation Type	Outer Cover	Special Provisions and Use
Moisture-resistant thermoplastic	TW	60 (140°F)	Flame-retardant Moisture-resistant Thermoplastic	None	Dry and wet locations
Heat-resistant thermoplastic	THHN	90 (194°F)	Flame-retardant Moisture-resistant Thermoplastic	Nylon jacket	Dry locations
Moisture and heat-resistant thermoplastic	THW	75 (167°F)	Flame-retardant, moisture and heat-resistant thermoplastic	Nylon jacket	Dry and wet locations
Moisture and heat-resistant thermoplastic	THWN	75 (167°F)	Flame-retardant, moisture and heat-resistant thermoplastic	Nylon jacket	Dry and wet locations
Mineral insulation (metal sheathed)	MI	85 (185°F)	Magnesium oxide	Copper	Dry-wet locations with Type O terminating fittings. Max. oper. temp. for special appl. 250°C.
Thermoplastic and asbestos	TA	90 (194°F)	Thermoplastic and asbestos	Flame-retardant cotton braid	Switchboard wiring only
Varnished cambric	V	85 (185°F)	Varnished cambric	Nonmetallic covering or lead sheath	Dry locations only
Asbestos and varnished cambric	AVA	110 (230°F)	Impregnated asbestos and varnished cambric	Asbestos braid	Dry locations only

Asbestos and varnished cambric	AVL	110 (230°F)	Impregnated asbestos and varnished cambric	Lead sheath	General use and wet locations
Asbestos and varnished cambric	AVB	90 (194°F)	Impregnated asbestos and varnished cambric	Flame-retardant cotton braid	Dry locations only
Asbestos	A	200 (392°F)	Asbestos	Without asbestos braid	Dry locations only. In raceways only for leads to or within apparatus limited to 300 V.
Asbestos	AA	200 (392°F)	Asbestos	With asbestos braid	Dry locations. Open wiring. In raceways for leads to or within apparatus limited to 300 V.
Asbestos	AI	125 (257°F)	Impregnated asbestos	Without asbestos braid	Dry locations only. In raceways only for leads to or within apparatus limited to 300 V.
Asbestos	AIA	125 (257°F)	Impregnated asbestos	With asbestos braid	Dry locations. Open wiring. Raceways for leads to or within apparatus.
Heat and moisture-resistant cross-linked polyethylene	XHHW	75 (167°F) 90 (194°F)	Cross-linked polyethylene	None	Wet and dry locations Dry locations
Moisture, heat and oil-resistant thermoplastic	THW-MTW	60 (140°F) 90 (194°F)	Thermoplastic	None	Wet and dry locations Special applications within electric discharge lighting equipment (size 14-8 only).

TABLE 7-11 DIMENSIONS OF ASBESTOS-VARNISHED-CAMBRIC INSULATED CONDUCTORS

Size AWG (MCM)	Type AVA Diam. (in.)	Type AVA Area (in.^{2})	Type AVB Diam. (in.)	Type AVB Area (in.^{2})	Type AVL Diam. (in.)	Type AVL Area (in.^{2})
14	.245	.047	.205	.033	.320	.080
12	.265	.055	.225	.040	.340	.091
10	.285	.064	.245	.047	.360	.102
8	.310	.075	.270	.057	.390	.119
6	.395	.122	.345	.094	.430	.145
4	.445	.155	.395	.123	.480	.181
2	.505	.200	.460	.166	.570	.255
1	.585	.268	.540	.229	.620	.300
0	.625	.307	.580	.264	.660	.341
00	.670	.353	.625	.307	.705	.390
000	.720	.406	.675	.358	.755	.447
0000	.780	.478	.735	.425	.815	.521
250	.885	.616	.855	.572	.955	.715
300	.940	.692	.910	.649	1.010	.800
350	.995	.778	.965	.731	1.060	.885
400	1.040	.850	1.010	.800	1.105	.960
500	1.125	.995	1.095	.945	1.190	1.118
550	1.165	1.065	1.135	1.01	1.265	1.26
600	1.205	1.140	1.175	1.09	1.305	1.34
650	1.240	1.21	1.210	1.15	1.340	1.41
700	1.275	1.28	1.245	1.22	1.375	1.49
750	1.310	1.35	1.280	1.29	1.410	1.57
800	1.345	1.42	1.315	1.36	1.440	1.63
850	1.375	1.49	1.345	1.43	1.470	1.70
900	1.405	1.55	1.375	1.49	1.505	1.78
950	1.435	1.62	1.405	1.55	1.535	1.85
1000	1.465	1.69	1.435	1.62	1.565	1.93

Notes: No. 14 to No. 8, solid; No. 6 and larger, stranded; except AVL, where all sizes are stranded.

Table includes information on Type AVA, AVB, and AVL insulated conductors.

Dimensions given for diameter and area are approximate.

8

Fuses

INTRODUCTION

All electrical circuits and their related components are subject to destructive overcurrent. Harsh environments, general deterioration, accidental damage or damage from natural causes, excessive expansion or overloading of the electrical system are factors that contribute to the occurrence of such overcurrents. Reliable protective devices prevent or minimize costly damage to transformers, conductors, motors, and the many other components and loads that make up the complete distribution system. Reliable circuit protection is essential to avoid severe monetary losses, which can result from power blackouts and prolonged downtime of facilities. To protect electrical conductors and equipment against abnormal operating conditions and their consequences, protective devices are used in circuits. The fuse is one such device.

Overcurrents

An overcurrent is either an overload current or a short-circuit current. The overload current is a current that is excessive relative to normal operating current but one which is confined to the normal conductive paths provided by the conductors and other components and loads of the electrical system.

A short circuit (Fig. 8-1) is probably the most common cause of electrical problems. It is an undesired current path that allows the elec-

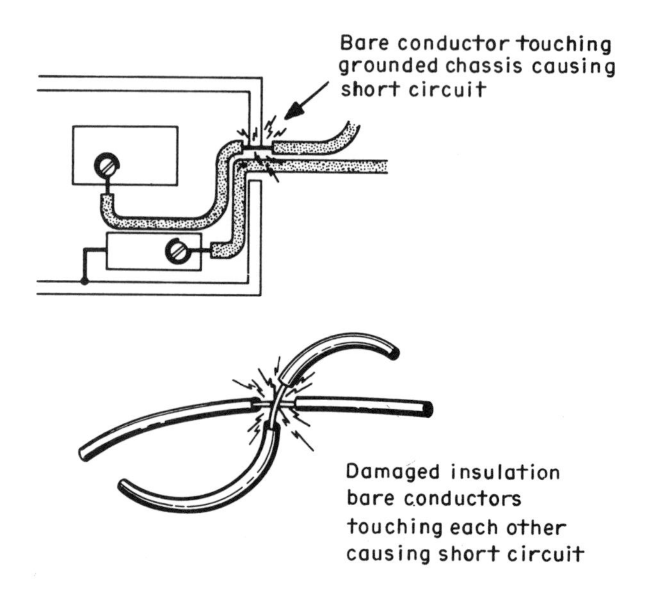

Figure 8-1 A short circuit is probably the most common cause of electrical problems.

trical current to bypass the load on the circuit. Sometimes the short is between two wires due to faulty insulation or it can occur between a wire and a grounded object.

Overloads

Overloads are most often between one and six times the normal current level. Usually, they are caused by harmless temporary surge currents that occur when motors are started up or transformers are energized. Such overload currents normally occur. Since they are of brief duration, any temperature rise is trivial and has no harmful effect on the circuit components. (It is important that protective devices do not react to them.)

Continuous overloads can result from defective motors (such as worn motor bearings), overloaded equipment, or too many loads on one circuit. Such sustained overloads are destructive and must be cut-off by protective devices before they damage the distribution system or system loads. However, since they are of relatively low magnitude compared to short-circuit currents, removal of the overload current within a few seconds will generally prevent equipment damage. A sustained overload current results in overheating of conductors and other components and will cause deterioration of insulation which may eventually result in severe damage and short circuits if not interrupted.

Definition

A fuse is the simplest device for opening an electric circuit when excessive current flows because of an overload or such fault conditions as grounds and short circuits. A "fusible" link or links encapsulated in a tube and connected to contact terminals comprise the fundamental elements of the basic fuse. Electrical resistance of the link is so low that it simply acts as a conductor, and every fuse is intended to be connected in series with each hot conductor so that current flowing through the conductor to any load must also pass through the fuse. The continuous current rating of the fuse in amperes establishes the maximum amount of current the fuse will carry without opening. When circuit current flow exceeds this value, an internal element (link) in the fuse melts due to the heat of the current flow and opens the circuit.

FUSE RATINGS

Fuses are made in a wide variety of types and sizes with different current ratings, different abilities to interrupt fault currents, various speeds of operation (either quick-opening or time-delay opening), different internal and external constructions, and voltage ratings for both low-voltage (600 V and below) and medium-voltage (over 600 V) circuits.

Voltage Rating

Most low voltage power distribution fuses have 250-V or 600-V ratings (other ratings are 125 V and 300 V). The voltage rating of a fuse must be at least equal to the circuit voltage. It can be higher but never lower. For instance, a 600-V fuse can be used in a 208-V circuit. The voltage rating of a fuse is a function of or depends upon its capability to open a circuit under an overcurrent condition. Specifically, the voltage rating determines the ability of the fuse to suppress the internal arcing that occurs after a fuse link melts and an arc is produced. If a fuse is used with a voltage rating lower than the circuit voltage, arc suppression will be impaired and, under some fault current conditions, the fuse may not safely clear the overcurrent.

Ampere Rating

Every fuse has a specific ampere rating. In selecting the ampacity of a fuse, consideration must be given to the type of load and code requirements. The ampere rating of a fuse should normally not exceed current carrying capacity of the circuit. For instance, if a conductor is rated to carry 20 A, a 20-A fuse is the largest that should be used in the conductor circuit. However, there are some specific circumstances when the ampere rating is permitted to be greater than the current carrying capacity of the circuit. A typical example is the motor circuit; dual-element fuses generally are permitted to be sized up to 175% and non-time-delay fuses up to 300% of the motor full-load amperes. Generally, the ampere rating of a fuse and switch combination should be selected at 125% of the load current (this corresponds to the circuit capacity which is also selected at 125% of the load current). There are exceptions, such as when the fuse-switch combination is approved for continuous operation at 100% of its rating.

A protective device must be able to withstand the destructive energy of short-circuited currents. If a fault current exceeds a level beyond the capability of the protective device, the device may actually rupture and cause severe damage. Thus, it is important in applying a fuse or circuit breaker to use one which can sustain the largest potential short-circuit currents. The rating that defines the capacity of a protective device to maintain its integrity when reacting to fault currents is termed its *interrupting rating*. The interrupting rating of most branch-circuit, molded case, circuit breakers typically used in residential service entrance boxes is 10,000 A. The rating is usually expressed as "10,000 amperes interrupting capacity (AIC)." Larger, more expensive circuit breakers may have AIC's of 14,000 A or higher. In contrast, most modern, current-limiting fuses have an interrupting capacity of 200,000 A and are commonly used to protect the lower rated circuit breakers. The NE Code, Section 110-9, requires equipment intended to break current at fault levels to have an interrupting rating sufficient for the current that must be interrupted.

Time Delay

The time-delay rating of a fuse is established by standard UL tests. All fuses have an inverse time current characteristic. That is, the fuse will open quickly on high currents and after a period of time delay, on low overcurrents. Specific types of fuses are made to have specially determined amounts of time delay. The basic UL requirement on time delay for Class RK-1, RK-15, and J fuses which are marked "time delay" is that the fuse must carry a current equal to five times its continuous rating for a period not less than ten sec. UL has not developed time-delay tests for all fuse classes. Fuses are available for use where time delay is needed along with current limitation on high-level short circuits. In all cases, manufacturers' literature should be consulted to determine the degree of time delay in relation to the operating characteristics of the circuit being protected.

TYPES OF FUSES

Plug Fuses

Plug fuses (Fig. 8-2) screw into fuseholders in the same manner that incandescent lamps screw into lampholders. There are two types of plug fuses: Edison-base and Type S. Edison-base plug fuses were the standard plug fuses used for many years. An Edison-base fuse consists of a strip of fusible (capable of being melted) metal in a small porcelain or glass case, with the fuse strip, or link, visible through a "window" in the top of the fuse. The screw base corresponds to the base of a standard medium-base incandescent lamp.

The chief disadvantage of the Edison-base plug fuse is that it is made in several ratings from 0 to 30 A, all with the same size base, permitting unsafe replacement of one rating by a higher rating. Today, the Edison-base plug fuse is recognized only for replacing existing Edison-base fuses. Section 240-51(b) of the NE Code says, "Plug fuses of the Edison-base type shall be used only for replacements in existing installations where there is no evidence of overfusing or tampering." For all new electrical installations where plug fuses are used, they must be Type S plug fuses.

Type S plug fuses were developed to reduce the possibility of overfusing a circuit (inserting a fuse with a rating greater than that required by the circuit). There are 15 classifications of Type S fuses based on current rating: 0–30 A. Each Type S fuse has a base of a different size and a matching adapter. Once an adapter is screwed into a standard Edison-base fuseholder, it locks into place and is not readily removed without destroying the fuseholder. As a result, only a Type S fuse with a size the same as that of the adapter may be inserted.

Plug fuses also are made in time-delay types that permit a longer period of overload flow before operation, such as on motor inrush current and other higher-than-normal rated currents. They are available in ratings up to 30 A, both in Edison-base and Type S. Their principal use is in motor circuits, where the starting inrush current to the motor is much higher than the running, or continuous, current. The time-

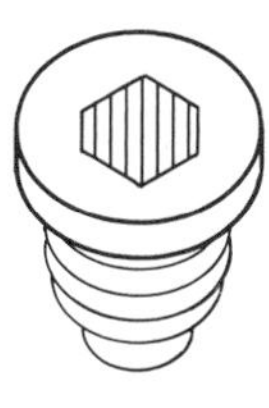

Figure 8-2 Typical plug fuse. Once blown, it must be replaced with a new fuse.

delay fuse will not open on the inrush of high-starting current. If, however, the high current persists, the fuse will open the circuit just as if a short circuit or heavy overload current had developed. All Type S fuses are time-delay fuses.

Plug fuses are permitted to be used in circuits of no more than 125 V between conductors, but they may be used where the voltage between any ungrounded conductor and ground is not more than 150 V. The screwshell of the fuseholders for plug fuses must be connected to the load side circuit conductor; the base contact is connected to the lineside or conductor supply. A disconnecting means (switch) is not required on the supply side of a plug fuse.

The plug fuse is a *nonrenewable* fuse; that is, once it has opened the circuit because of a fault or overload, it cannot be used again or renewed. It must be replaced by a new fuse of the same rating and characteristics for safe and effective restoration of circuit operation.

Single-Element Cartridge Fuses

The basic component of a fuse is the link. Depending upon the ampere rating of the fuse, the single-element fuse may have one or more links. They are electrically connected to the end blades (or ferrules) and enclosed in a tube or cartridge surrounded by an arc-quenching filler material (Fig. 8-3).

Under normal operation, when the fuse is operating at or near its ampere rating, it simply functions as a conductor. However, if an overload current occurs and persists for more than a short interval of time, the temperature of the link eventually reaches a level that causes a restricted segment of the link to melt; as a result, a gap is formed and an electric arc established. However, as the arc causes the link metal to burn back, the gap becomes progressively larger. Electrical resistance of the arc eventually reaches such a high level that the arc cannot be sustained and is extinguished; the fuse will have then completely cut off all current flow in the circuit. Suppression or quenching of the arc is accelerated by the filler material.

Single-element fuses of present day design have a very high speed

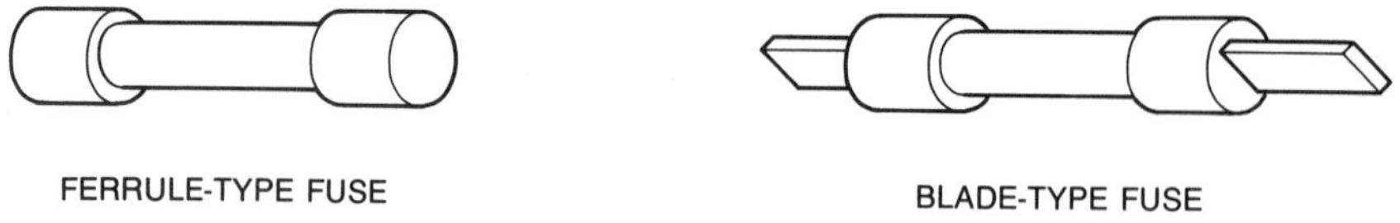

FERRULE-TYPE FUSE

BLADE-TYPE FUSE

Figure 8-3 Dual-element plug fuses hold when motors start, yet open quickly when shorts occur. They are available in 15, 20, 25, and 30 A.

of response to overcurrents. They provide excellent short-circuit component protection. However, temporary harmless overloads or surge currents may cause nuisance openings unless these fuses are oversized. They are best used, therefore, in circuits not subject to heavy transient surge currents and the temporary overload of circuits with inductive loads such as motors, transformers, and solenoids. Because single-element fuses have a high speed-of-response to short-circuit currents, they are particularly suited for the protection of circuit breakers with low interrupting ratings.

Whereas an overload current normally falls within the region of between one and six times normal current, short-circuit currents are quite high. The fuse may be subjected to short-circuit currents of 30,000 or 40,000 A or higher. Response of current-limiting fuses to such currents is extremely fast. The restricted sections of the fuse link will simultaneously melt (within a matter of two- or three-thousandths of a second in the event of a high-level fault current). The high total resistance of the multiple arcs together with the quenching effects of the filler particles results in rapid arc suppression and clearing of the circuit. Short-circuit current is cut-off in less than a half-cycle, long before the short-circuit current can reach its full value (fuse operating in its current limiting range).

Dual-Element Cartridge Fuses

Unlike single-element fuses, the dual-element fuse can be applied in circuits subject to temporary motor overloads and surge currents to provide both high performance short-circuit and overload protection. Oversizing in order to prevent nuisance openings is not necessary. The dual-element fuse contains two distinctly separate types of elements. Electrically, the two elements are series connected. The fuse links similar to those used in the single-element fuse perform the short-circuit protection function; the overload element provides protection against low-level overcurrents or overloads and will hold an overload that is five times greater than the ampere rating of the fuse for a minimum time of ten sec.

The overload section consists of a copper heat absorber and a spring-operated trigger assembly. The heat-absorber strip is permanently connected to the short-circuit link and to the short-circuit lii. on the opposite end of the fuse by the *S*-shaped connector of the trigger assembly. The connector electronically joins the one short-circuit link to the heat absorber in the overload section of the fuse. These elements are joined by a "calibrated" fusing alloy. An overload current causes heating of the short-circuit link connected to the trigger assembly. Transfer of heat from the short-circuit link to the heat absorbing strip in the midsection of the fuse begins to raise the temperature of the heat absorber. If the overload is sustained, the temperature of the heat absorber eventually reaches a level that permits the trigger spring to "fracture" the calibrated fusing alloy and pull the connector free. The short-circuit link is electrically disconnected from the heat absorber, the conducting path through the fuse is opened, and overload current is interrupted. A critical aspect of the fusing alloy is that it retains its original characteristic after repeated temporary overloads without degradation.

Dual-element fuses may also be used in circuits other than motor branch circuits and feeders, such as lighting circuits and those feeding mixed lighting and power loads. The low-resistance construction of the fuses offers cooler operation of the equipment, which permits higher loading of fuses in switch and panel enclosures without heat damage and without nuisance openings from accumulated ambient heat.

For years, the standard interruption or short-circuit rating of cartridge fuses was taken as 10,000 A. That is, if 10,000 A or less flowed through a fuse at the instant of short circuit, damage would not be transmitted outside the fuse. However, as electrical distribution systems and installations increased in size, available short-circuit currents increased far beyond 10,000 A. Often, when a short occurred and a fuse opened, switches or motor starters were damaged by the high momentary current that flowed, and sometimes the fuses actually blew apart. It was not unusual for switchgear busbars to be bent out of shape by the magnetic forces that were developed by the fault current. The need to prevent such high currents from flowing led to the development of the current-limiting fuse, a fuse with silver links surrounded by quartz sand. Such a fuse operates so quickly on a short circuit that the current wave has no time to increase to a damaging magnitude. Any fuse that opens the circuit in less than 1/2 cycle—that is, in less than 1/120 of a second—actually limits the current to a value less than that which would flow if the fuse were not in the circuit. Although all fuses provide some measure of current limitation if the fault current gets high enough, the term *current limiting* as a fuse characteristic is reserved for those fuses that satisfy stringent UL requirements. Today, there are several types of current-limiting fuses, each suited for a particular application.

FUSE MARKING

It is a requirement of the NE Code that cartridge fuses used for branch-circuit or feeder protection must be plainly marked, either by printing

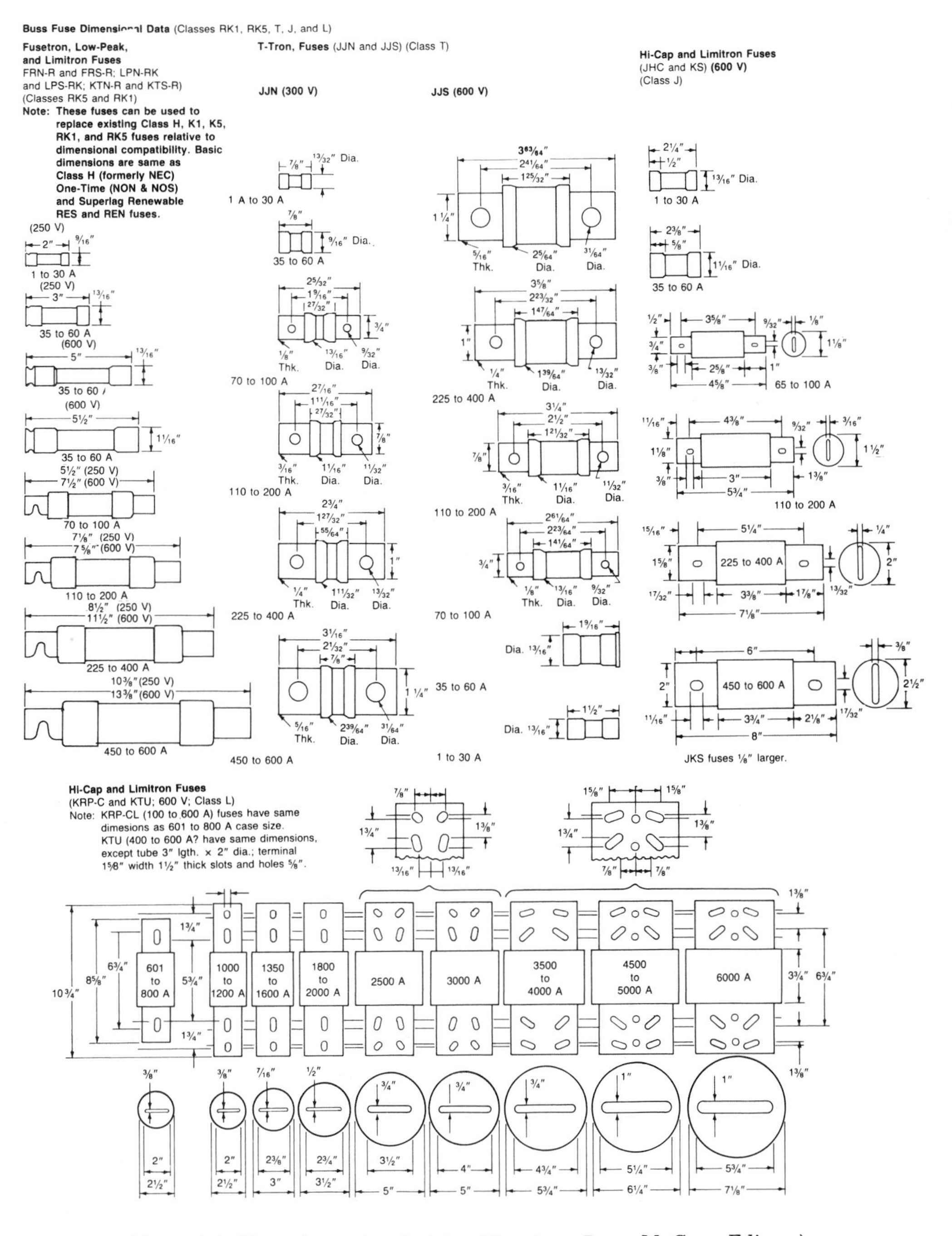

Figure 8-4 Fuse dimensional data. (Courtesy Buss, McGraw-Edison.)

on the fuse barrel or by a label attached to the barrel, showing the following:

1. ampere rating
2. voltage rating
3. interrupting rating (if other than 10,000 A)
4. "current limiting," where applicable
5. the name or trademark of the manufacturer

See Fig. 8-4 for fuse dimensions.

UNDERWRITERS' LABORATORIES' FUSE CLASSES

Fuses are tested and listed by Underwriters' Laboratories Inc. in accordance with established standards of construction and performance. There are many varieties of miscellaneous fuses used for special purposes or for supplementary protection of individual types of electrical equipment. However, here we will chiefly be concerned with those fuses used for protection of branch circuits and feeders on systems operating at 600 V or below. Plug fuses have been covered above. Cartridge fuses in this category include UL Class H, K, G, J, R, T, CC, and L. Each will be discussed in turn.

Class H Fuses

The Class H fuse is the type long referred to as the "standard NE Code fuse." It is the only type available with renewable links. It was originally manufactured in accordance with NEMA standards, but today all new Class H fuses are constructed to UL standards. If the available short-circuit current is over 10,000 A, Class H fuses should be replaced by Class K or R, which have higher interrupting ratings. Class R fuses will fit in Class H fuseholders. One word of caution: Some NEMA Class H fuses were constructed with short-circuit ratings of 200,000 A. If such a fuse needs replacement, the available short-circuit current should be determined and a Class K or R fuse used that has the appropriate interrupting rating. A UL Class H fuse should not be used in that case.

While UL Class H fuses are tested in a circuit with an available short-circuit current of 10,000 A, UL does not officially recognize an interrupting rating for Class H fuses. Therefore, they should be used only if the available short-circuit current is 10,000 A or less.

Although renewable Class H fuses offer some time delay, the

renewable link cannot be designed to meet the UL requirement of a 10-sec delay when carrying 500% of rated current. Class H fuses are available in ratings from 0 to 600 A at either 250 or 600 V. The Class H fuse is not considered a current-limiting fuse.

Class K Fuses

These nonrenewable fuses were developed to provide interrupting capacities in excess of 10,000 A. They are made in three classes: K-1, K-5, and K-9. One of these three designations appears on the fuse label; the fuse is never labeled Class K. Class K-1, K-5, and K-9 fuses have interrupting ratings of 50,000; 100,000; and 200,000 A, respectively. All three classes are current limiting. Class K-1 is most current limiting, Class K-9 is least current limiting, and Class K-5 is somewhere between.

Class K-1, K-5, and K-9 fuses may be obtained with time-delay characteristics and so marked provided that they meet the UL requirement of a 10-sec delay at 500% rated current. All Class K fuses are available in current ratings from 0 to 600 A at either 250 or 600 V. They have the same dimensions as Class H fuses and thus will fit in Class H fuseholders.

Class G Fuses

These are small-dimension cartridge fuses rated at 300 V to ground. Available in four different case sizes with ratings of 15, 20, 30, and 60 A, these fuses may be used in circuits rated up to 300 V between ungrounded conductors and on 480/277-V systems (where the voltage to ground does not exceed 300 V). Time delay is optional. The UL requirement for the Class G fuse is a 12-sec delay at 200% rated current. The fuse offers fast, current-limiting action in its operation on short circuits. These characteristics, combined with an interrupting rating of 100,000 A, make the Class G fuse suitable for use on commercial and industrial systems where very high short-circuit currents exist. These fuses are often used in equipment where a size rejection (as with the Type S plug fuse) branch circuit fuse is required.

Class J Fuses

The Class J fuse is a nonreversible, current-limiting fuse with an interrupting rating of 200,000 A. It is available in ratings of 0 to 600 A at 600 V and is not interchangeable with any other UL fuse class. UL standards do provide for time delay for Class J fuses. Class J fuses were once manufactured in accordance with NEMA standards. When it becomes necessary to replace a NEMA Class J fuse, a UL Class J fuse or its equivalent should be used. Class J fuses rated over 60 A have drilled blades to permit optional bolting in place.

Class R Fuses

These are recently developed current-limiting fuses with special rejection features. The overall dimensions of the Class R fuses are the same as Class H or K fuses, and therefore Class R fuses can be used as replacements for Class H or K fuses. However, the Class R fuse has a notched ferrule or notched blade and is used with special rejection-type fuseholders that will not accept Class H fuses, which are not current limited. When equipment is protected by current-limiting fuses, the bus structures, conductors, starters, and other electrical components can be selected with less bracing than if noncurrent-limiting fuses or standard circuit breakers were used. The rejection fuseholders prevent the replacement of current-limiting fuses with noncurrent-limiting types that would not limit the current and therefore would not constitute adequate equipment protection. This potential hazard is recognized by the NE Code in Section 240-60(b). Class R fuses (0 to 600 A, 250, and 600 V), have an interrupting rating of 200,000 A. They are marked either RK-1 or RK-5. These are subclassifications of Class R fuses to further recognize their degree of current limitation. The Class RK-1 fuse is the most current limiting, with characteristics very similar to the Class K-1 fuse; the Class RK-5 fuse is slightly less current limiting, with characteristics very similar to the Class K-5 fuse.

Class T Fuses

These are relatively new current-limiting fuses (0 to 1200 A, 300, and 600 V) with small physical dimensions and a 200,000-A interrupting rating. Their small size makes it possible to reduce greatly the size of equipment in which they are used. Fuseholders designed to accept Class T fuses will reject all other fuse classes. No time-delay tests have been established. The 300-V fuse may be used on solidly grounded 277/480-V systems.

Class CC Fuses

The Class CC is a new, small, nonrenewable fuse for control circuits and intended to make possible a reduction in the size of enclosure used. Ratings are 0 to 30 A, 600 V, and 200,000 A interrupting rating. The Class CC fuse is not interchangeable with any other fuse class in equipment provided with Class CC fuseholders. No time-delay tests have been established for this fuse.

Class L Fuses

These are nonrenewable, current-limiting, high-capacity fuses available in ratings from 60 to 6000 A, 600 V, with an interrupting rating of 200,000 A. Class L fuses were previously manufactured in accordance with NEMA standards, and when NEMA Class L fuses need replacing, UL Class L fuses should be used. These fuses are not interchangeable with any other fuse class. Their contact blades are drilled for bolting onto busbars to assure positive contact.

Although UL does not test Class L fuses for time delay, they are manufactured with some time delay and are permitted to be labeled time delay. Comparing fuse characteristic curves of different manufacturers will indicate the degree of delay provided by each. High-capacity Class L fuses are well suited for use as main and feeder protection, since they can be coordinated easily to prevent nuisance outages and offer a high degree of current limitation for protection of the system components.

INTERCHANGING FUSES

When interchanging fuses, the most important consideration is that a fuse have an interrupting rating and current-limiting ability suitable for the circuit or system in which it is used. In general, a fuse with a higher interrupting rating and current-limiting ability may be used in place of a fuse with a lower interrupting rating and less current-limiting ability, but not vice versa. However, there is another consideration in making such substitutions. Assume that an old, Class H-fused switch needs replacements, and Class K-5 fuses are used instead of the Class H to provide increased interrupting capability. If the available short-circuit current has increased substantially over the years, the new fuses may handle a short circuit adequately, but the switch may blow up. Thus, a need for higher fuse interrupting ratings may also indicate a need for higher interrupting ratings for switches and other equipment.

The greatest hazard exists between Class H and Class K fuses, since Class H fuses will fit in Class K fuseholders. Class R fuses will fit in Class H and Class K fuseholders, but the rejection features of Class R fuseholders prevent other than the intended fuses from being inserted in them. In general, problems involving incorrect interchanging of fuses can only be solved by properly educating the user.

GUIDE FOR SIZING FUSES

General guidelines for sizing fuses are given here for most circuits that will be encountered on conventional systems. Some specific applications may warrant other fuse sizing; in these cases, the load characteristics and appropriate NE Code sections should be considered. The selections shown here are not, in all cases, the maximum or minimum ampere ratings permitted by the NE Code. Demand factors permitted per the NE Code are not included here.

Dual-Element Time-Delay Fuses

1. *Main service:* Each ungrounded service entrance conductor shall have a fuse in series with a rating not higher than the ampacity of the conductor, except as permitted in Article 230-90(a-d) of the NE Code. The service fuses shall be part of the service disconnecting means or be located immediately adjacent thereto (Article 230-91).
2. *Feeder circuit with no motor load:* The fuse size must be at least 125% of the continuous load plus 100% of the noncontinuous load. Do not size larger than ampacity of conductor.
3. *Feeder circuit with all motor loads:* Size the fuse at 150% of the full load current of the largest motor plus the full-load current of all motors.
4. *Feeder circuit with mixed loads:* Size fuse at sum of 150% of the full-load current of the largest motor plus 100% of the full-load current of all other motors plus 125% of the continuous, nonmotor load plus 100% of the noncontinuous, nonmotor load.
5. *Branch circuit with no motor load:* The fuse size must be at least 125% of the continuous load plus 100% of the noncontinuous load. Do not size larger than ampacity of conductor.
6. *Motor branch circuit with overload relays:* Where overload relays are sized for motor running overload protection, the following provide backup, ground fault, and short-circuit protection:
   A. Motor 1.15 service factor or 40°C rise: Size fuse at 125% of motor full-load current or next higher standard size.
   B. Motor less than 1.15 service factor or over 40°C rise: Size the fuse at 115% of the motor full-load current or the next higher standard fuse size.
7. *Motor branch circuit with fuse protection only:* Where the fuse is the only motor protection, the following fuses provide motor running overload protection and short-circuit protection.

A. Motor 1.15 service factor or 40°C rise: Size the fuse at 100% to 125% of the motor full-load current.
B. Motor less than 1.15 service factor or over 40°C rise: Size fuse at 100% to 115% of motor full-load current.

8. *Large motor branch circuit:* Fuse larger than 600 A. For large motors, size KRP-C HI-CAP time-delay fuse at 150% to 225% of the motor full-load current, depending on the starting method; that is, part-winding starting, reduced voltage starting, and so on.

Non-Time-Delay Fuses

1. *Main service:* Each ungrounded service entrance conductor shall have a fuse in series with a rating not higher than the ampacity of the conductor, except as permitted in Article 230-90(a-d) of the NE Code. The service fuses shall be part of the service disconnecting means or be located immediately adjacent thereto (Article 230-91).
2. *Feeder circuit with no motor loads:* The fuse size must be at least 125% of the continuous load plus 100% of the noncontinuous load. Do not size larger than the ampacity of the wire.
3. *Feeder circuit with all motor loads:* Size the fuse at 300% of the full-load current of the largest motor plus the full-load current of all other motors.
4. *Feeder circuit with mixed loads:* Size fuse at sum of 300% of full-load current of largest motor plus 100% of full-load current of all other motors plus 125% of the continuous, nonmotor load plus 100% of noncontinuous, nonmotor load.
5. *Branch circuit with no motor load:* The fuse size must be at least 125% of the continuous load plus 100% of the noncontinuous load. Do not size larger than the ampacity of conductor.
6. *Motor branch circuit with overload relays:* Size the fuse as close to but not exceeding 300% of the motor running full-load current. Provides ground fault and short-circuit protection only.
7. *Motor branch circuit with fuse protection only:* Non-time-delay fuses cannot be sized close enough to provide motor running overload protection. If sized for motor overload protection, non-time-delay fuses would open due to the motor starting current. Use dual-element fuses.

When sizing fuses for a given application, a schematic drawing of the system will help tremendously. Such drawings do not have to be detailed, just a single-line schemaic, such as the one shown in Fig. 8-5 will suffice.

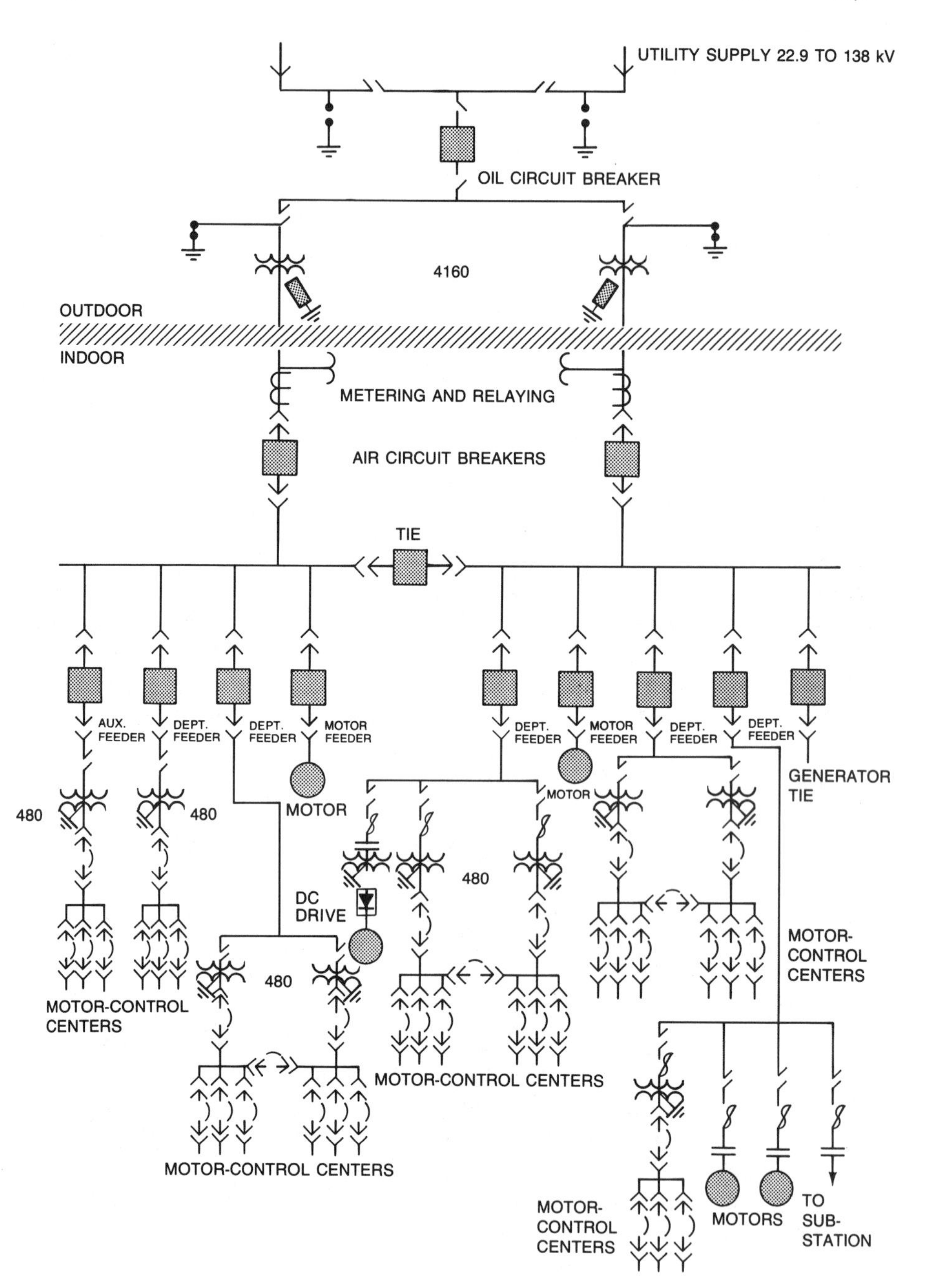

Figure 8-5 Schematic diagram of typical electrical distribution system.

COORDINATION

Coordination is the name given to the time-current relationship among a number of overcurrent devices connected in series. Examples include fuses in a main feeder, subfeeder, and branch circuits. Safety is the prime consideration in the operation of fuses; however, coordination of the characteristics of fuses has become a very important factor in the large and complex electrical system of the present time.

Every fuse should be properly rated for continuous current and overloads and for the maximum short-circuit current the electrical system could feed into a fault on the load side of the fuse—but this is not enough! It might still be possible for a fault on a feeder to open the main service fuse before the feeder fuse opens. Or, a branch-circuit fault might open the feeder fuse before the branch-circuit device opens. Such applications are said to be uncoordinated, or nonselective—the fuse closest to the fault is not faster operating than one farther from the fault. When a fuse on a feeder opens the main service fuse instead of the feeder fuse, all of the electrical system is taken out of service instead of just the one faulted feeder. Effective coordination minimizes the extent of electrical outage when a fault occurs. It therefore minimizes loss of production, interruption of critical continuous processes, loss of vital facilities, and possible panic.

Selective coordination is the selection of overcurrent devices with time/current characteristics that assure clearing of a fault or short circuit by the device nearest the fault on the line side of the fault. A fault on a branch circuit is cleared by the branch-circuit device. The subfeeder, feeder, and main service over-current devices will not operate. Or, a fault on a feeder is opened by the feeder fuse without opening any other fuse on the supply side of the feeder. With selective coordination, only the faulted part of the system is taken out of service, which represents the condition of minimum outage.

With proper selective coordination, every device is rated for the maximum fault current it might be called upon to open. Coordination is achieved by studying the curve of current versus time required for operation of each device. Selection is then made so that the device nearest any load is faster operating than all devices closer to the supply, and each device going back to the service entrance is faster operating than all devices closer to the supply. The main service fuses must have the longest opening time for any branch or feeder fault.

Speed of fuse operation sometimes makes it difficult to coordinate fuses with a circuit breaker on its load side. However, the opposite—a circuit breaker with fuses on the load side—may offer some coordination, with the fuse operating before the circuit breaker for faults on the load side of the fuse. This depends on fuse size, circuit breaker setting, and the like. Fuse manufacturers make available coordination data on the use of their fuses, including the ratio of sizes in which a fuse will operate before larger fuses with the same or different operating characteristics. Data are also provided for coordinating fuses with circuit breakers.

9

Circuit Breakers and Switches

The NE Code requires overcurrent protection at the main service entrance equipment and also for individual circuits to protect the overall electrical installation, or parts of it, against ground faults, overloads, and electrical surges of all types.

At one time, the main service disconnecting device consisted of a main switch capable of disconnecting the entire installation with one throw of a handle. Fuses were installed in the switch to protect the electrical system as a whole. The main switch usually fed a fuse cabinet of four or more circuits. This fuse cabinet provided overcurrent protection for individual branch circuits. Figure 9-1 shows a power-riser diagram of such a service installation. Figure 9-2 shows an old type of service-entrance equipment; it contains a main fuse block, a fuse block for the electric range, and four fuse holders for branch circuits.

Although fusible safety switches are still used considerably for disconnecting and overload protection for motors and similar pieces of electrical equipment, the majority of service entrances utilize main distribution centers, with either fusible or circuit breaker overload protective devices. However, safety switches are used at times on some commercial and industrial applications, for service-entrance equipment, mainly in specialized situations where the time and expense required to have a special main distribution panel constructed does not warrant its use.

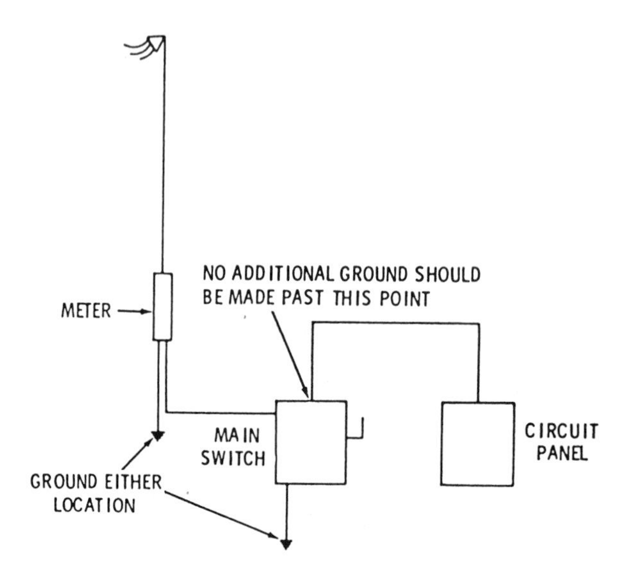

Figure 9-1 Power-riser diagram of a fused disconnect service entrance.

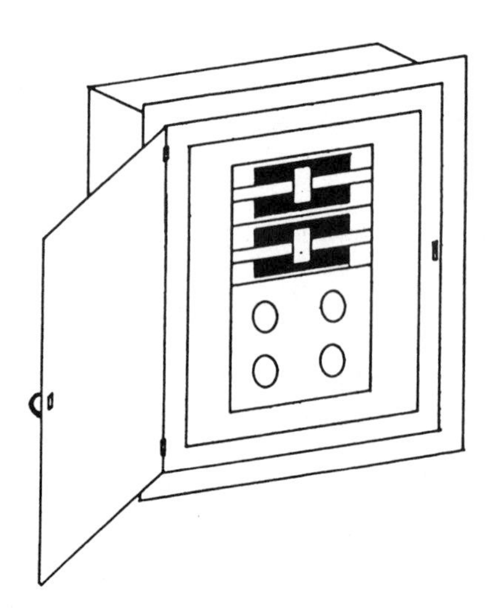

Figure 9-2 Fuse panel that is seldom used any more.

FUSIBLE SERVICE EQUIPMENT

Fusible safety switches, or fuse blocks, are rated at 30, 60, 100, 200, 400, 600, A, and so on and all are usually readily available for distributing to electrical contractors, manufacturing plants, and others. Although the switches themselves do not have any in-between ratings, the fuses that are installed may be any rating at or below the rating of the switch itself. For example, a situation may require a disconnect switch

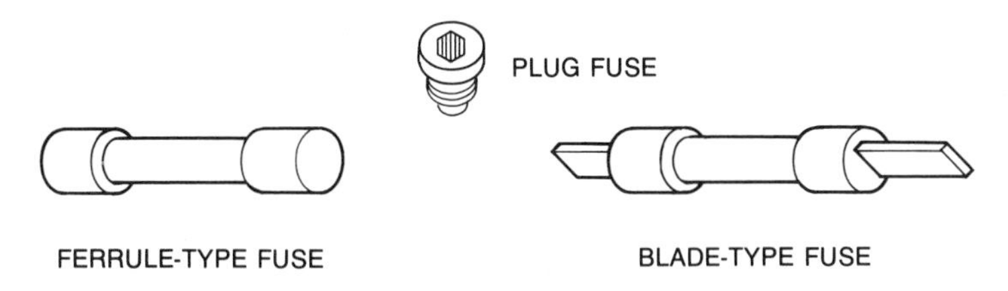

Figure 9-3 Three of the most common types of cartridge fuses.

with an overcurrent rating of 150 A. In this case, a 200-A fusible safety switch will have to be installed, but fused at 150 A. The wiring may also be rated at only 150 A in this case.

Fusible main distribution panels are also available that contain fuse block for both single and three-phase circuits. They resemble circuit-breaker MDP's (Main Distribution Panel) in appearance and utilize approximately the same amount of space.

Fusible panelboards are available, with or without main fuse blocks, and up to 40 spaces for branch circuits. One fuse is needed for each 120- or 277-V circuit, two fuses are required for two- and three-wire, single-phase systems, while three fuses are used for all three-phase circuits.

Plug fuses are sometimes used for 120/240-V systems for up to 30-A ratings. If the circuits require ratings of over 30 A, cartridge fuses are necessary; three of the most common types are shown in Fig. 9-3.

Any switch used for electrical construction should be approved and labeled by Underwriters' Laboratories, Inc., because some cheap switches utilized in the past have been known to cause injury.

CIRCUIT BREAKERS

A circuit breaker resembles an ordinary toggle switch, and it is probably the most widely used means of overcurrent protection today. On an overload, the circuit breaker opens itself or "trips." In a tripped position, the handle jumps to the middle position (Fig. 9-4). To reset, turn the handle to the Off position and then turn it as far as it will go beyond this position; finally turn it to the On position.

One single-pole breaker protects a 120-V circuit; and one double-pole breaker protects a 240-V circuit; while three-pole circuit breakers are used for three-wire, three-phase circuits. The breakers are rated in amperes just like fuses, although the particular ratings are not exactly the same as those for fuses.

Circuit-breaker enclosures come in several types; one contains only branch-circuit breakers, while another contains a main-circuit

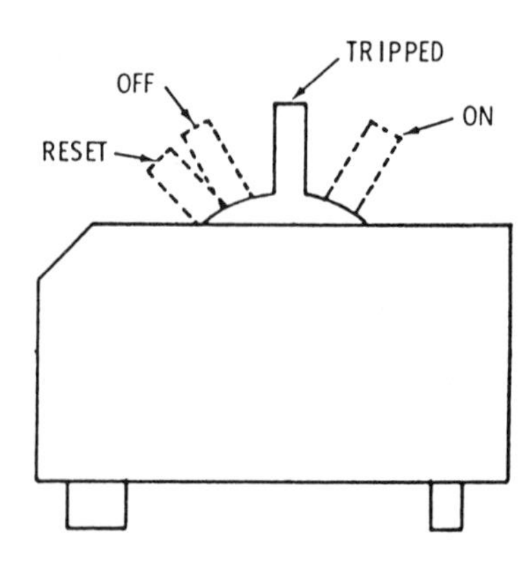

Figure 9-4 Operating characteristics of circuit breaker.

breaker in addition to branch-circuit breakers. Most of the types used for residential applications are of the plug-in type; that is, the cabinets are usually sold without the breakers but contain an arrangement of busbars. The user then selects whatever combination of breakers he requires and plugs them into this busbar arrangement. Most industrial and commercial circuit-breaker panels utilize bolt-in type circuit-breakers, and heavy-duty enclosures. All types will be thoroughly explored in this chapter.

If the circuit breaker cabinet contains six or less circuit breakers, it is permissible to eliminate a main-disconnecting means, provided the breakers are rated at more than 20 A.

Two types of automatic overload devices normally used in electrical circuits to prevent fires or the destruction of the circuit and its associated equipment are fuses and circuit breakers.

Basically, a circuit breaker is a device for closing and interrupting a circuit between separable contacts under both normal and abnormal conditions. This is done manually (normal condition) by use of its handle by switching to the On or Off positions. However, the circuit breaker also is designed to open a circuit automatically on a predetermined overload or short-circuit current without damage to itself or its associated equipment. As long as a circuit breaker is applied within its rating, it will automatically interrupt any short and therefore must be classified as an inherently safe, overcurrent protective device. The molded case housing made of phenolic plastic of the circuit breaker insulates all live parts, thereby protecting the associated equipment on the circuit as well as operating personnel from accidental contact.

A standard molded case circuit breaker usually contains

1. A set of contacts
2. A magnetic trip element
3. A thermal trip element

4. Line and load terminals
5. Bussing used to connect these individual parts
6. An enclosing housing of insulating material

The circuit-breaker handle manually opens and closes the contacts and resets the automatic trip units after an interruption. Some circuit breakers also contain a manually operated "push-to-trip" testing mechanism.

Circuit breakers are grouped for identification according to given current ranges. Each group is classified by the largest ampere rating of its range. These groups are: 15-100 A, 125-225 A, 250-400 A, 500-1000 A, and 1200-2000 A. Therefore, they are classified as 100, 225, 400, 1000, and 2000 A frames. These numbers are commonly referred to as *frame classifications* or *frame sizes* and are the terms applied to groups of molded case circuit breakers that are physically interchangeable with each other.

Voltage Rating

The established voltage rating of a circuit breaker is based on its clearances or space, both through air and over surfaces between all components of the electrical circuit and between the electrical components and ground. Circuit breaker voltage ratings indicate the maximum electrical system voltage on which they can be applied.

A circuit breaker can be rated for either alternating current (ac) or direct current (dc) system applications or for both. Single-pole circuit breakers, rated at 120/240-V ac or 125/250-V dc can be used singly and in pairs on three-wire circuits having a neutral connected to the midpoint of the load. Single-pole circuit breakers rated at 120/240-V ac or 125/250-V dc also can be used in pairs on a two-wire circuit connected to the outside (ungrounded) wires of a three-wire system. Two-pole or three-pole circuit breakers rated 120/240-V ac or 125/250-V dc can be used only on a three-wire, direct current, or single-phase, alternating current system having a grounded neutral. Circuit-breaker voltage ratings must be equal to or greater than voltage of the electrical system on which they are used.

Circuit breakers have two types of current ratings. The first—and the one that is used most often—is the continuous-current rating. The second is the short circuit current interrupting capacity.

Current Rating

The rated continuous current of a device is the maximum current in amperes that it will continuously carry without exceeding the specified limits of observable temperature rise. Continuous-current ratings of circuit breakers are established based on standard UL ampere ratings. These are 15, 20, 25, 30, 35, 40, 45, 50, 60, 70, 80, 90, 100, 110, 125, 150, 175, 200, 225, 250, 300, 350, 400, 450, 500, 600, 700, 800, 1000, 1200, 1600, 2000, 2500, 3000, 4000, 5000, and 6000 A. The ampere rating of a circuit breaker is located on the handle of the device and the numerical value alone is shown.

General application requires that the circuit breaker current rating must be equal to or less than the load circuit conductor current-carrying capacity (ampacity).

Interrupting Capacity Rating

The AIC rating of a circuit breaker is the maximum short-circuit current that the breaker will interrupt safely. This AIC rating is at rated voltage and frequency.

A circuit breaker must be selected with interrupting capacity equal to or greater than the available short circuit current at the point where the circuit breaker is applied in the system. The breaker interrupting capacity is based on tests to which the breaker is subjected. There are two such tests; one set up by UL and the other by NEMA. The NEMA tests are self-certification while UL tests are certified by unbiased witnesses. UL tests had been limited to a maximum of 10,000 A in the past, so emphasis was placed on NEMA tests with higher ratings. UL tests now include the NEMA tests plus other ratings so that now the emphasis is being placed on UL tests.

The interrupting capacity of a circuit breaker is based on its rated voltage. Where the circuit breaker can be used on more than one voltage, the interrupting capacity will be shown for each voltage level. For example, the LA-type circuit breaker has 42,000 A, symmetrical interrupting capacity at 240 V, 30,000 A symmetrical at 480 V, and 22,000 A symmetrical at 600 V.

Standard Interrupting Capacity

Standard interrupting-capacity-rated circuit breakers can be identified by the black operating handle and black printed interrupting rating labels.

The interrupting rating of a circuit breaker is as important in application as the voltage and current ratings and should be considered each time a breaker is applied.

High Interrupting Capacity (I-75,000 breakers)

Where higher interrupting capacity than the standard ratings is required in the 15A–100-A FA-type circuit breaker, the FH (*H* for high interrupting capacity) is available. The FH, classified as an I-

75,000-type circuit breaker, has an interrupting rating of 65,000 A, symmetrical at 240-V ac. The continuous-current ratings are duplicated in these breakers (15A–100 A) but the interrupting capacity has been increased to satisfy the need for greater interrupting capacity. This type of breaker is applied in installations where the higher short-circuit currents are available (such as large industrial plants).

I-75,000-type circuit breakers are identified quickly by their gray colored handles and red interrupting rating labels. This is in contrast to the black handles on breakers with standard interrupting ratings. The I-75,000 breakers are dimensionally identical to the lower interrupting capacity rated breakers, but are physically different. They are built with a case material that will withstand higher shocks from heat and interrupting forces and they are extremely flame resistant. I-75,000 breakers are available in the FH, KH, LH, MH, NH, and PH types.

Three ampere ratings (15, 20, and 30) are also available in the QH (I-75,000) type breaker. These breakers satisfy conditions of lighting circuits supplied from these high available-short-circuit-current systems.

The FA-type circuit breaker has been used as an example but the discussion is applicable to all I-75,000 types of molded case circuit breakers.

Current-Limiting Circuit Breakers

The need for current limitation is the result of increasingly higher available short-circuit currents associated with the growth and interconnection of modern power systems. To meet this need, engineers have had to face difficult design decisions.

Until now, these decisions have always had to involve some kind of compromise. One could not take advantage of the versatility and convenience of a molded case circuit breaker and at the same time retain the current-limiting downstream protection of a current-limiting fuse.

This dilemma can now be solved by utilizing the I-Limiter—a practical current-limiting circuit breaker. The I-Limiter operates so fast and limits so well that it can provide downstream protection for other Square D branch breakers with as little as 10,000 AIC rating on systems with 100,000 A available fault current. The I-Limiter does not use fuses. There is nothing to replace—even after clearing maximum-level fault currents.

Molded Case Switch

NE Code Article 430-102 requires a disconnecting means be placed within sight of all motor-control locations. Nonautomatic circuit interrupters are used to meet this requirement when the overcurrent protection is out of sight from the controller. The nonautomatic interrupter is a device without trip elements. It consists of the standard breaker contacts, bussing, and lugs for the highest ampere rating in each breaker frame size and is solely manually operated. Although these devices are primarily used in motor-branch circuits, they need not be horsepower rated because their interrupting ability is much greater than the locked rotor currents produced on such circuits. Nonautomatic interrupters will withstand fault current levels between 15 and 20 times the continuous rating. FA, KA, LA, MA, and PA nonautomatic circuit interrupters are available. These devices are not available in the I-75,000 type (except for the PE breaker).

Automatic molded case switches are used where the available fault current is greater than the nonautomatic interrupter will withstand and is generally a safer device to use than the nonautomatic interrupter. This device has a fixed magnetic-trip element set to actuate on current values equal to or greater than 20 times the breaker rating. This automatic-circuit interrupter is not intended to be a circuit overcurrent protective device. High setting of the trip element allows the feeder to branch, circuit overcurrent protective device to do its job without interference. The automatic-circuit interrupter has the same interrupting rating as the thermal-magnet circuit breakers in its group. FA, FH, KA, KH, LA, LH, MA, NH, PA, PH, and PC type circuit interrupters are available. This includes both the standard- and I-75,000-interrupting ratings.

Terminal Lugs

All circuit breakers have provisions on both the line and load sides for making connection in the electrical circuit. These connections can be plug-on, bolt-on, solderless lugs, wire binder screws, or crimp-type lugs.

Auxiliary Devices

The advantage of the circuit breaker goes beyond permitting the disconnection and protection of the electrical circuit. By adding auxiliary devices, a breaker can do such things as sense voltage on its own system or other systems, warn personnel of faulty conditions, disconnect or control associated equipment on completely different types of electrical systems as well as mechanical devices, and allow itself to be controlled from remote locations along with many other operations.

A shunt trip is a mechanism that trips a circuit breaker by means of a trip coil energized from a separate circuit or source of power. The trip-coil circuit is closed by a relay, switch, or other means. The shunt trip is available in two- and three-pole breakers and is normally installed in the left pole. Shunt trip coils do not have a continuous-current rating. A cut-off switch is included to break the coil circuit when the breaker opens. Standard shunt trips are rated 6, 12, 24, 48, 125, and 250-V dc; and 120, 208, 240, 277, 480, and 600-V ac. Other ratings are available upon special request. The control leads for the shunt trip are color coded black.

An undervoltage trip device is one that trips a circuit breaker automatically when the main circuit voltage decreases to approximately 40% of its value. The breaker cannot be reset until the voltage returns to 80% of normal value. These trips are available in two- and three-pole breakers and are installed in the left pole. They are supplied as standard in the same voltage ratings as the shunt trip. Trips rated above 24-V ac are supplied with external resistors. The undervoltage trip is not supplied with time-delay action. The control wires for undervoltage trips are color coded black. Since the shunt trip and undervoltage trip is normally furnished in the same pole of the circuit breaker, only the undervoltage trip is necessary to obtain the operation of both devices when they are installed on the same electrical system. A normally closed contact, such as those used in stop buttons, can be installed in the control circuit to open the breaker in a manner similar to a shunt trip.

An auxiliary switch is one that is mechanically operated by the main switching device for signalling, interlocking, or other purposes. An A-type contact is one which is open when the breaker contacts are open. The B-type contact is closed when the breaker contacts are open. Auxiliary switches are available in two- and three-pole breakers and are normally installed in the right pole.

The alarm switch is a single-pole device that is activated when the breaker is in the tripped position. It is used to actuate bell alarms and warning signals. The alarm switch is factory installed and is rated at least one ampere at 120-V ac.

ENCLOSURES

The majority of circuit breakers and fuses are used in some types of enclosure (e.g., panelboards, switchboards, motor control centers, and individual enclosures).

NEMA has established enclosure designations because individually enclosed circuit breakers are used in so many different types of locations, weather and water conditions, and dust and other contaminating conditions. The designation (e.g., NEMA 12) indicates an enclosure type to fulfill requirements for a particular application. The

NEMA designations were recently revised to obtain a clearer and more precise definition of the enclosure needed to meet various standard requirements.

Some of the revisions in the NEMA designations, which are of interest to us in this section are

- The NEMA Type 1A (semi-dust tight) has been dropped.
- The NEMA 12 enclosure can now be substituted in many installations in place of the NEMA 5.
- The advantage of this substitution is that the NEMA 12 enclosure is much less expensive than the NEMA 5 enclosure.
- NEMA Type 3R as applied to circuit breaker enclosures is a lighter weight, less expensive rainproof enclosure than the other weather resistant enclosure types.

Types of enclosures generally available are as follows. For other types, the factory should be consulted.

NEMA 1: A general purpose enclosure primarily intended to prevent accidental contact with the enclosed parts. It is suitable for general purpose applications indoors where it is not exposed to unusual service conditions.

NEMA 1A: A general purpose enclosure with gaskets (now obsolete). NEMA 12 is now used.

NEMA 3R: A rainproof enclosure primarily intended to protect against a beating rain. It is suitable for general applications outdoors where sleet-proof construction is not required.

NEMA 4 and 5: A watertight enclosure designed to exclude water applied in the form of a hose stream. Will exclude dust.

NEMA 5: Dust-tight. A combination enclosure for both 4 and 5 is often listed.

NEMA 7: An enclosure used where a combustible vapor exists. It meets requirements for NE Code, Class I, Division 1, Group D. (Examples of this atmosphere are: gasoline vapors, natural gas, paint vapors, and flammable cleaning fluids.)

NEMA 9: An enclosure used where combustible dust exists. It meets requirements for NE Code, Class II, Division 1, Groups E, F, and G. (Examples of this atmosphere are: flour, and aluminum, coal, and grain dust.)

NEMA 12: An industrial enclosure designed for use in those industries where it is desired to exclude such material as dust, lint, fibers and flyings, and oil and coolant seepage. These enclosures are available with and without knockouts.

NEUTRALS

An insulated neutral is provided with all Q1, (100 A) NEMA 1 and 3R enclosures used on 240-V systems. Bonding screws and straps are provided, but not installed, where the enclosure may be used as a service entrance device. Neutrals are available for installation in all other enclosures but are merchandised separately at ampere ratings of 100, 225, 400, 600, 800, and 1000.

SAFETY SWITCHES

Two types of safety switches are in common use: heavy duty and general duty, which both have visible blades and safety handles. With visible blades the contact blades are in full view so you can clearly see you're safe. Safety handles are always in complete control of the switch blades, so whether the cover is open or closed, when the handle is in the Off position the switch is always Off.

Heavy-duty switches are intended for applications where price is secondary to safety and continued performance. This type of switch is usually subjected to frequent operation and rough handling. Heavy-duty switches are also used in atmospheres where a general-duty switch would be unsuitable. Heavy duty switches are widely used by automobile manufacturers, breweries, foundries, ship yards, and similar heavy industries. Most heavy duty switches are rated 30 A through 1200 A, 240 V to 600 V (ac or dc). The switches with horsepower ratings are able to interrupt approximately six times the full-load, motor-current ratings. When equipped with Class J or Class R fuses, heavy duty safety switches are UL listed for use on systems with up to 200,000 A available fault current, RMS symmetrical. This is the highest withstand rating of the industry.

Heavy-duty switches are available with NEMA 1, 3R, 4, 4X, 5, 7, 9, and 12 enclosures.

General-duty switches are for residential and light commercial applications where the price of the device is a limiting factor. General-duty switches are meant to be used where operation and handling are moderate and where the available fault current is less than 10,000 A. Some examples of general duty switch applications would be: residential service entrance, light duty branch circuit disconnects, major appliance disconnects, and farm and small business service entrances.

General-duty switches are rated up to 600 A at 240 V (ac only) in general purpose (NEMA 1) and rainproof (NEMA 3R) enclosures. These switches are horsepower rated and capable of opening a circuit with approximately six times a motor's full-load current rating.

All current-carrying parts of general duty switches are plated with either tin or cadmium to reduce heating. Switch jaws and blades are made of plated copper for high conductivity. A steel reinforcing spring increases the mechanical strength of the jaws and assures a firm contact pressure between blade and jaw. High-pressure action means minimum contact resistance and cool operation. All general-duty switch blades feature *Visible Blade* design, so there is no doubt when the switch is Off.

Most general-duty safety switches are furnished with mechanical set-screw lugs suitable for terminations of either aluminum or copper conductors. Crimp-type lugs are seldom available on general-duty safety switches.

Double-Throw Safety Switches

Double-throw switches are used as transfer switches and are not intended as motor circuit switches, therefore they are not horsepower rated. Three lines of double-throw switches are in common use: 82,000 line and 92,000 line, plus the DTU rainproof manual-transfer switch.

The 82,000 line switches are available as either fused or unfused devices. These switches feature quick-make, quick-break action, plated current-carrying parts, a key-controlled interlock mechanism and screw-type lugs. Arc suppressors are supplied on all switches rated above 250 V. Provisions for up to three padlocks are available to lock the handle in either On or Off positions. The 82,000 line switches come in NEMA 1 and 3R enclosures.

The 92,000 line switches are manually operable and are not quick-make, quick-break. They are available as either fused or unfused devices in NEMA 1 enclosures only.

GROUND-FAULT PROTECTION FOR PEOPLE

Ground-fault protection for people is a subject of interest to all of us, both personally and professionally. A ground fault exists when an unintended path is established between an ungrounded conductor and ground. This situation can occur not only from worn or defective electrical equipment but also from accidental misuse of equipment that is in good working order.

Effects of Current on the Human Body

Hand-to-hand body resistance of an adult lies between 1000 Ω and 4000 Ω, depending on moisture, muscular structure, and voltage. The average value is 2100 Ω at 240-V ac and 2000 Ω at 120-V ac.

Using Ohm's law, the current resulting from the above average hand-to-hand resistance values is 114 ma (.114 A) at 240-V ac and 43 ma (.043 A) at 120-V ac. The effects of 60 Hz alternating current on a normal healthy adult are as follows (note that current is in milliamperes):

- More than 5 ma—generally painful shock.
- More than 15 ma—sufficient to cause "freezing" to the circulation for 50% of the population.
- More than 30 ma—breathing difficult (possible suffocation).
- 50 to 100 ma—possible ventricular fibrillation.
- 100 to 200 ma—certain ventricular fibrillation.
- Over 200 ma—severe burns, muscle contractions. The heart is more likely to stop than fibrillate.

The current that would flow from a defective electric drill, for example, through the metal housing and through the human body to ground would be 43 ma, calculated using 2800 Ω as average body resistance. Using 1000 Ω as body resistance, the current flow would be 120 ma.

Forty-three milliamperes is only 0.29% of the current required to open a 15-A circuit breaker or fuse, and yet it approaches the current level that may produce ventricular fibrillation. Obviously, the standard circuit breaker or fuse will not open the circuit under such low levels of current flow.

Ground Fault Circuit Interrupters (GFCI)

"People protector" devices are built as Class A devices in accordance with UL Standard No. 943 for Ground Fault Circuit Interrupters. UL defines a Class A device as one that "will trip when a fault current to ground is 6 milliamperes or more." The tripping time of such units cannot exceed the value obtained by the equation,

$$T = \left(\frac{20}{I}\right)1.43$$

where T is time in seconds and I is the ground-fault current in milliamperes. Also, Class A devices must not trip below 4 ma.

Class A GFCIs provide a self-contained means of testing the ground-fault circuitry, as required by UL. To test, simply push the test button and the device will respond with a trip indication. UL requires that the current generated by the test circuit shall not exceed 9 ma.

Also, UL requires the device to be functional at 85% of the rated voltage.

The GFCI sensor continuously monitors the current balance in the ungrounded "hot" load conductor and the neutral load conductor. If the current in the neutral load wire becomes less than the current in the hot load wire, then a ground fault exists, since a portion of the current is returning to the source by some means other than the neutral load wire. When an imbalance in current occurs, the sensor, which is a differential-current transformer, sends a signal to the solid state circuitry, which activates the ground trip solenoid mechanism and breaks the hot load connection. A current imbalance as low as 6 ma will cause the circuit breaker to interrupt the circuit. This will be indicated by the red VISI-TRIP indicator as well as the position of the operating handle centered between Off and On.

10

Panelboards

By definition, a *panelboard* or panel is a single panel or group of panel units designed for assembly in the form of a single panel that includes buses, automatic overcurrent devices, and often switches for the control of light, heat, or power circuits. The interior assemblies are designed to be placed in a cabinet or cutout box placed in or against a wall or partition and accessible only from the front.

Along the same lines is a *switchboard*, which consists of a large single panel, frame, or assembly of panels on which are mounted on the face or back, or both, switches, overcurrent and other protective devices, buses, and usually instruments. Unlike panelboards, switchboards are generally accessible from the rear as well as from the front and are not intended to be installed in cabinets.

Persons engaged in work in the electrical industry will also come across the term *load center*, which is the name normally given panelboards for use as residential, mobile home, and apartment building distribution equipment.

MAIN-LUGS LOAD CENTERS

Main-lugs load centers provide distribution of electrical power where a main disconnect with overcurrent protection is provided separately from the load center. All terminals on modern equipment are suitable for both aluminum or copper conductors.

Most main-lugs load centers, 125 A and up, have interiors which are reversible for either top or bottom feed. The cover does not need to be reversed when the interior is reversed, as the cover will fit the housing and interior components in either position. Most single-phase, three-wire load centers are also approved for three-phase grounded *B* systems and 240-V ac. A main-lugs load center can also be converted to a main-breaker load center very easily by installing a main circuit breaker onto the busbars and then back feeding that breaker with the service or feeder conductors. After the main breaker is inserted, the number of spaces available for branch circuits is equal to the number of original spaces in the load center, less the number of spaces taken up by the main breaker.

A 600-A rated load center is available for garden, townhouse, and other types of two- to six-unit apartment complexes where individual metering is not required.

MAIN-BREAKER LOAD CENTERS

Main-breaker load centers have many time-saving features that can save electrical contractors considerable expense. First of all, the main breaker is factory installed, cutting installation costs. There are no lugs to remove, no screws, no nuts, or washers to misplace, and no expensive main breaker to become misplaced. Factory assembled main disconnects also assure a proper and safe electrical connection. The main breaker and neutral terminals are located at the same end of the load center and are adjacent to one another, which allows straight-in wiring and eliminates awkward bends and space-cramping loops in the incoming service cables. Also, since the neutral location is not in the branch wiring gutter, the load gutters carry only branch-breaker connections and are not cluttered with neutral or ground conductors.

Housings 14 in. wide are available in 100 A–225 A main-breaker load centers, which offers more side gutter space for wiring, and permits flush load center installation between 16-in. centered studs without an extra mounting support to hold the box in place.

The neutral bars for branch circuits have alternating lugs rated No. 14-8 and No. 14-4 wire size, listed only for one conductor per hole, and are suitable for copper or aluminum conductors. The line-side terminals of the main breakers are also suitable for use with copper or aluminum conductors.

Doors on main breaker load centers are designed to completely cover both the branch breakers and the main breaker, leaving no exposed breaker handles to detract from the appearance of the load center or to be accidentally switched.

SPLIT-BUS LOAD CENTERS

The NE Code allows the use of a maximum of six main disconnects in a common enclosure. Split-bus load centers have the bus split or divided into sections, which are insulated from each other to provide an economical service entrance device in applications not requiring a single main disconnect.

The main section of split-bus load centers has provisions for up to six main disconnects for the heavier 240-V appliances, subfeeders and lighting main disconnects. The lower section contains provisions for lighting and 120-V appliance circuits and is fed by the lighting main disconnect, which is located in the main bus section. All split-bus devices are provided with factory-installed wires connected to the lower section. These wires can then be field connected to the lighting main disconnect in the main section.

Some of the most recent designs have provisions for a field installable lighting main using a plug-on breaker 60 A–125 A. Any of these breakers can be used without changing the factory installed lighting main wire described previously. A 14-in. wide housing is standard on this type of load center to allow installation between studs in partitions.

RISER PANELS

Riser panels, consisting of a main-lugs only load center with an extended gutter of over 6 in., are ideally suited for high-rise office buildings and for certain apartment complexes. Most are available with 6, 8, and 12 circuit load centers. The box (housing), interior and covers are sold separately so they can be installed at the most convenient time during construction, which reduces the possibility of loss and minimizes the number of parts to stock.

Another type of riser panel is the feed-through load center. In this panel, the main busbars have lugs at both ends and therefore actually become part of the riser system. When using these in high-rise buildings, the savings in riser wire length needed can be considerable. The branch breakers merely plug onto the main busbars. Feed-through riser panels do not have a main disconnect.

A standard load center may be converted in the field to a riser panel by adding one of the appropriate auxiliary gutters, which may be attached to either the right or the left side of the load center. The auxiliary gutters and the tap kits are sold separately, thereby saving space in the warehouse and adding even more versatility to the load center line of equipment.

DISTRIBUTION EQUIPMENT

Panelboard manufacturers offer a complete line of lighting and distribution panelboards, most of which are available either unassembled from distributor stock or factory assembled.

NQO panelboards are rated for use on the following ac services:

- 120/240-V, single-phase, three-wire
- 240-V, three-phase, three-wire delta
- 240-V, three-phase delta with grounded B-phase
- 120/208-V, three-phase, four-wire wye

They carry no dc rating. NQO panelboards are available either factory assembled or unassembled.

This type of panelboard is suitable for use in industrial buildings, schools, office and commercial buildings, and institutions when the largest branch breaker does not exceed 150 A and the system voltage is not greater than 240-V ac.

NQO panelboards have maximum-mains ratings of 400-A main breaker or main lugs. Branch circuit breakers may be catalog prefix QO, QO-H, QH, Q1, or Q1-H; one-, two-, or three-pole; have a maximum rating of 150 A; and feature plug-on bus connections. QO and Q1 circuit breakers are standard with 10,000-AIC rating and QH breakers with 65,000-AIC rating. Other ratings for specific applications are also available.

Qwik-Gard branch circuit breakers with ground-fault circuit interruption may also be supplied in type NQO panelboards. Rated 10,000-AIC symmetrical, these GFCI devices provide UL Class A (5 ma sensitivity) ground-fault protection as well as overload and short-circuit protection for branch-circuit wiring.

NQO unassembled panelboards are available as follows:

1. branch breakers
2. interior with solid neutral
3. box, either 14 in. wide × 4 in. deep, 14 in. wide × 5-3/4 in. deep or 20 in. wide × 5-3/4 in. deep
4. mono-flat front with door and flush lock
5. accessories

NQO factory-assembled panelboards are identical in construction to the unassembled type. Mains ratings and branch circuits are the same. Unlike unassembled panelboards, however, the branches are factory installed.

In NQO construction both assembled and unassembled boxes are fabricated of galvanized steel. A variety of knockouts is provided in each end wall. Interiors having maximum 225-A main-lugs rating are available in 14 in. wide × 4 in. deep; 14 in. wide × 5-3/4 in. deep or 20 in. wide × 5-3/4 in. deep boxes. Boxes 14 in. wide × 5-3/4 in. deep or 20 in. wide × 5-3/4 in. deep are required for panelboards having a main circuit breaker. Boxes for interiors having 400-A mains (breakers or lugs) are 20 in. wide × 5-3/4 in. deep.

Interiors for standard width panelboards having a maximum rating of 225 A are of the *single bus* construction. In this construction, one-, two-, and three-pole catalog prefix Q1 breakers extend the full width of the panelboard and cannot be mounted opposite each other. QO, QO-H, and QH circuit breakers twin mount on the bus assembly. In other words, a three pole Q1 requires six QO pole spaces.

Interiors rated at a maximum 400-A utilize a *double-row bus* construction. This type of construction consists of two sets of busbars mounted on a single pan. The respective phase buses of each set are paralleled with each other by means of insulated, solid connectors. QO and Q1 breakers mount on a one-for-one basis (i.e., a three pole QO requires the same spaces as a three pole Q1).

All current-carrying parts are plated for maximum corrosion resistance and minimum heating at contact surfaces. Main lugs are UL listed for use with either copper or aluminum cable. Main lugs may be replaced by the appropriate Anderson type VCEL crimp lug, when required. Lug catalog numbers and crimp tool type are called out on the panelboard wiring diagram. Box-type lugs for circuits on both branch breakers and the solid neutral permit maximum convenience and speed in wiring and also are UL listed for use with either copper or aluminum cable.

Mono-Flat fronts are constructed of one-piece steel finished in gray baked enamel. Door hinges and adjustable trim clamps are completely concealed. After the panelboard door is locked, the front cannot be removed—a desirable feature when used in schools, commercial and industrial buildings and institutions. With the Mono-Flat front, it is extremely easy to either paint or wallpaper over the trim to match the wall covering. These fronts are available for flush or surface mounting.

Many additional features are available on factory-assembled panelboards. Increased mains, split buses, sub-feed lugs and breakers, lighting contactors, and time clocks are among the special features available. Weatherproof and dustproof enclosures for specific applications, and finishes other than standard gray (USAS #49) are also available on factory-assembled panelboards.

COLUMN-WIDTH PANELBOARDS

In industrial buildings, the only place to centrally locate lighting panelboards is often within the web of a structural steel column. Column-width panelboards are frequently used with cable trough and a pull box for maximum wiring space. The neutral connections are made in the pull box to reduce the number of wires in the necessarily narrow cable trough (wireway). The number of wires in the cable trough is governed by limitations of NE Code Section 362-5. For such installations, column-width NQO panelboards are available. Boxes 8-5/8 in. wide are designed for installations in 10-in. wide flange (WF) structural steel columns and boxes 6-7/8 in. wide are designed for installation in 8-in.-WF structural columns.

Column-width panelboard interiors are similar in construction to standard-width interiors, with the exception that the circuit breakers are mounted in a single vertical row rather than twin mounted. The size of the wiring gutter limits the type of circuit breaker to QO only (Q1 breaker-branch conductors require more than 2-in. wiring gutter). Due to space limitations, the Mono-Flat front cannot be furnished on either the 6-7/8-in. or the 8-5/8-in. column width panelboards. The front uses conventional hinges and is mounted with external screws.

NQO column-width panelboards are available unassembled in the 8-5/8-in. wide × 5-in. deep box size.

TYPE NQOB PANELBOARDS

NQOB panelboards have service voltage ratings identical to those of Type NQO and are available either unassembled or factory assembled. Branch breakers are bolted in place in NQOB panelboards, rather than plugged into the busbars. NQOB panelboards are suitable for use in industrial buildings, schools, office and commercial buildings and institutions when the largest branch breaker does not exceed 100 A and the system voltage is not greater than 240-V ac.

Branch circuit breakers must be of the QOB, QOB-H, QHB, Q1B, or Q1B-H catalog prefix. All branch breakers are available in either one-, two-, or three-pole construction; are rated 15 A–100 A and are bolt-on type (indicated by a *B*, e.g., QOB). Bolt-on Qwik-Gard circuit breakers with ground fault circuit interruption as described for plug-on under Type NQO panelboards are also available in Type NQOB panelboards.

The NQOB panelboard is furnished where the specifying authority requires bolted connections rather than plug-on connections.

METER CENTERS

Another form of panelboard that has grown in application is the meter center panelboard. Meter centers consist of prewired hookups of electric energy (kilowatt-hours) meter sockets and circuit disconnect devices (with overcurrent protection) for the circuits derived from the center, all combined in one enclosure package. Small units for individual residences consist of one socket in combination with service entrance panel facilities for the branch circuits in the building. Larger assemblies consist of a number of meter enclosures with disconnect and protection (such as a circuit breaker) for each of the subcircuits broken down from the main circuit supply to the meter center.

Use of meter distribution centers has been spurred by the continued construction of buildings with individual tenants who require separate metering such as in apartment houses, shopping centers, office buildings, and professional buildings. The prewired hookup of each meter and its disconnect device to the main busbars fed by the supply circuit to the meter center offers substantial installation economy to electrical contractors as long as use of such meter centers is permitted by the local electric power company. These centers eliminate the nippling together of individual components and the detailed wiring together, using separate meter enclosures, control and protective devices, and gutters.

PANELBOARD GRAPHICS AND SCHEDULES

All electrical sales personnel should be familiar with the various symbols and schedules used on electrical working drawings. Although the exact method of describing these items on working drawings may vary considerably, the following examples should acquaint the reader with the general procedures.

The panelboard schedule in Fig. 10-1 is typical of those used on electrical working drawings to give pertinent data on the service panelboards within a building. This schedule provides sufficient data to identify the panel number (as will more than likely be shown on the accompanying working drawings), the type of cabinet (either surface-mounted or flush), the panel main busbars or circuit breaker (in amperes, volts, and phase), or both, the number and type of circuit breakers contained in the panelboard, and the items fed by each. This type of schedule, however, does not give detailed information concerning the individual circuits, such as the wire size or the number of outlets on the circuit: this latter information must be given elsewhere on the drawing, usually in the plan view or in power-riser diagrams.

The draftsman draws the schedule on working drawings, using a straightedge and triangles, measuring off the various dimensions with the architect's scale. Guidelines are then made with the Ames lettering guide and the information is finally lettered (freehand or with mechanical lettering).

A practical application of this form is shown in Fig. 10-2. Here, the panelboard type is identified by the letter *A*. The location of this panelboard within the building will be shown on the floor plans by the appropriate symbol and identified again as panel *A*. The schedule indicates that the panel cabinet is to be surface mounted and is to contain a 200-A, 120/240-V, single-phase (three-wire) main circuit breaker. The panelboard is to be manufactured by Square D and is to be Type NQOB.

The column under the heading *Branches* gives data pertaining to the overcurrent protection devices for the individual branch circuits. This schedule indicates that the panel will contain fourteen 1-pole circuit breakers (for 120-V loads) with a trip rating of 20 A to furnish overcurrent protection for the lighting and receptacle circuits. The schedule also calls for six 2-pole circuit breakers rated at 20 A to furnish overcurrent protection for the lighting and receptacle circuits. The schedule also calls for six 2-pole circuit breakers rated at 20 A to furnish overcurrent protection for the air conditioning units and motors; one 2-pole, 50-A circuit breaker for the kitchen unit (combination cooktop and refrigerator); one 2-pole, 40-A circuit breaker for the time switch controlling the outside lights; six 1-pole circuit breakers as spares for later electrical additions; and two blank spaces for future provisions also.

The wire size is not indicated on this schedule, but the electrician installing the system will know that a 20-A circuit requires No. 12 AWG wire, a 30-A circuit requires No. 10 AWG wire, and so on. The wire sizes usually are indicated on the plan drawing also.

Figure 10-3 shows another type of panelboard schedule used on another project. This type not only has spaces for the data on the former schedule, but also has provisions for numbering the circuits so that the designer can balance the load. This schedule gives information that includes the total load on each circuit (in voltamperes or watts), and the size and type of overcurrent protection. The space at the bottom of the panel is for the manufacturer and type of panelboard to be installed as well as for providing other necessary information.

PANEL BOARD SCHEDULE											
PANEL NO.	TYPE CABINET	PANEL MAINS			BRANCHES					ITEM FED OR REMARKS
		AMPS	VOLTS	PHASE	1P	2P	3P	PROT	FRAME	

Figure 10-1 Typical panelboard schedule used on working drawings.

PANEL BOARD SCHEDULE										
PANEL NO.	TYPE CABINET	PANEL MAINS			BRANCHES					ITEM FED OR REMARKS
		AMPS	VOLTS	PHASE	1P	2P	3P	PROT	FRAME	
A	SURFACE	600	120/208 V	3ø, 4-W	21	—	—	20	70	LIGHTS
SQUARE D TYPE NQOB W/MAIN CIRCUIT BREAKER					12	—	—	20	70	RECEPTS
					—	—	/	30	70	TIME CLOCK
					—	—	/	200	200	PANEL G
					—	—	/	200	200	PANEL P
P	SURFACE	200	120/208 V	3ø, 4-W						
SQUARE D TYPE NQOB FOR HTG AND COOLING EQUIP.										

Figure 10-2 Practical application of the form shown in Fig. 10-1.

PANEL____ ______ 1ϕ3 WIRE________ MOUNTED_______ AMPERE MAIN____________

LOCATION______________________________ _______ AMPERE BUS

CCT.	VOLT – AMPERES			OUTLETS			CCT BKR		PHASE		CCT BKR		OUTLETS				VOLT – AMPERES		CCT.
NO.	ϕA	ϕB	DESCRIPTION		LTG	REC	POLE	TA	A	B	TA	POLE	REC	LTG		DESCRIPTION	ϕA	ϕB	NO.
1									●										2
3										●									4
5									●										6
7										●									8
9									●										10
11										●									12
13									●										14
15										●									16
17									●										18
19										●									20
21									●										22
23										●									24
25									●										26
27										●									28
29									●										30
31										●									32
33									●										34
35										●									36
37									●										38
39										●									40
								SUB – TOTALS											
			TOTAL VA/ϕ																
			LCL ADDER																
			TOTAL VA																
			LINE AMPS																

Figure 10-3 An alternate type of panelboard schedule used on working drawing.

PANELBOARD SPECIFICATIONS

In most cases it is the electrical contractor's responsibility to interpret working drawings and written specifications that are prepared by consulting engineering firms. The following is a typical electrical specification, in condensed form, from an actual electrical specification:

PANELBOARDS

a. General: Furnish and install circuit-breaker panelboards as indicated in the panelboard schedule and where shown on the drawings. The panelboard shall be dead front safety type equipped with molded case circuit breakers and shall be the type as listed in the panelboard schedule. Service entrance panelboards shall include a full capacity box bonding strap and approved for service entrance. The acceptable manufacturers of the panelboards are ______________________ provided that they are fully equal to the type listed on the drawings. The panelboard shall be listed by Underwriters' Laboratories and bear the UL label.

b. Circuit breakers: Provide molded case circuit breakers of frame, trip rating, and interrupting capacity as shown on the schedule. Also, provide the number of spaces for future circuit breakers as shown in the schedule. The circuit breakers shall be quick-make, quick-break, thermal-magnetic, trip indicating and have common trip on all multipole breakers with internal tie mechanism.

c. Panelboard bus assembly: Busbar connections to the branch circuit breakers shall be the *phase sequence* type. Single-phase, three-wire panelboard busing shall be such that any two adjacent single-pole breakers are connected to opposite polarities in such a manner that two-pole breakers can be installed in any location. Three-phase, four-wire breakers are individually connected to each of the three different phases in such a manner that two- or three-pole breakers can be installed at any location. All current-carrying parts of the bus assembly shall be plated. Mains ratings shall be as shown in the panelboard schedule on the plans. Provide solid neutral (S/N) assembly when required.

d. Wiring terminals: Terminals for feeder conductors to the panelboard mains and neutral shall be suitable for the type of conductor specified. Terminals for branch circuit wiring, both breaker and neutral, shall be suitable for the type of conductor specified.

e. Cabinets and fronts: The panelboard bus assembly shall be enclosed in a steel cabinet. The size of the wiring gutters and gauge of steel shall be in accordance with NEMA Standards. The box shall be fabricated from galvanized steel or equivalent rust-resistant steel. Fronts shall include door and have flush, brushed stainless steel, spring-loaded door pulls. The flush lock shall not protrude beyond the front of the door. All panelboard locks shall be keyed alike. Fronts shall not be removable with door in the locked position.

f. Directory: On the inside of the door of each cabinet, provide a typewritten directory, which will indicate the location of the equipment or outlets supplied by each circuit. The directory shall be mounted on a metal frame with a nonbreakable transparent cover. The panelboard designation shall be typed on the directory card and panel designation stenciled in 1-1/2-in. high letters on the inside of the door.

g. Panelboard installation

1. Before installing panelboards check all of the architectural drawings for possible conflict of space and adjust the location of the panelboard to prevent such conflict with other items.
2. When the panelboard is recessed into a wall serving an area with accessible ceiling space, provide and install an empty conduit system for future wiring. All 1-1/4-in. conduit shall be stubbed into the ceiling space above the panelboard and under the panelboard if such accessible ceiling space exists.
3. The panelboards shall be mounted in accordance with Article 373 of the NE Code. The electrical contractor shall furnish all material for mounting the panelboards.

11

Bus and Trolley Duct

INTRODUCTION

Various types and sizes of trolley and bus duct systems with self-contained conductors are manufactured to be used as a substitute for conduit and wire or cable for specialized feeder and power installations. Such systems are installed exposed with the proper hangers after the building construction is completed.

A *busway* consists of a metal enclosure within which rigid conductors, usually rectangular copper or aluminum busbars, are mounted on insulating supports. Branch circuit busways are usually installed in industrial plants to provide flexibility in connecting lighting fixtures and small electric tools.

A complete line of plug-in and feeder busway is available to produce a highly flexible and efficient power system. Most systems are available in ratings from 600 A through 4000 A (aluminum), and from 800 A through 5000 A (copper). Most are constructed in three-pole or four-pole full neutral and for system voltages up to 600-V ac. They may carry two-pole ratings with UL listing up to 8650 A dc or single-phase ac.

In general, busway may be thought of as prewired raceway—a wiring method which is an alternative to using individual single-conductor cables in conduit. As an alternative to using approved cables or insulated conductors in conduit, busway has a higher material cost; but because the need for installing conductors in the housing on the job is eliminated and because other work details involving termination and tapping of standard conductors is replaced by fast, standard techniques, the installation labor for busway is greatly reduced, making it an effective and often extremely advantageous alternative to other wiring methods.

Use of busway has been a fast growing trend in the electrification of modern commercial and industrial buildings. Busway is very popular for high-capacity risers, which are vertical feeders carrying power to the various floors of high-rise office and apartment buildings and the like. In both commercial and industrial buildings, such as convention halls and factories, busways are widely used for both vertical and horizontal power distribution—either as circuits directly to large load devices or as feeders to switchboard or panelboard centers where the load is broken down to smaller circuits.

The engineering design of busway systems is often considerably less involved than design of wire and conduit distribution circuits. Because busways provide packaged distribution systems of factory-engineered capacity and characteristics, they greatly simplify layout of large, industrial-type distribution systems. And because busway systems are assembled on a "building-block" basis, they go together faster and easier and can be dismantled or rearranged with the same speed and ease. This feature particularly adapts them to use in factories where the need for periodic rearrangement of production lines or other operations can be met by the layout flexibility of busway, minimizing disruption of production. In plants where supply-circuit flexibility is very important, busway is frequently the only economic solution.

Other design and application advantages of busway derive from the well-established and tested electrical specs provided by the manufacturers. Voltage-drop characteristics, for instance, are fixed and known and the data provided for a given busway simplifies calculations. The tested-and-known impedance ratings and the established short-circuit rating of busway, which tell the user how much current the busway can handle under fault conditions, enable the user to evaluate a given length of busway for application at any point in a system. Still another advantage is the range of sizes, which permits the user to readily select a busway of suitable rating for the known load and the anticipated amount of load growth, thus providing a way to know the exact amount of spare capacity. All of these features add up to a safer, more effective, more economical application in many cases.

TOTALLY ENCLOSED BUSWAYS

Many feeder busway and plug-in busway are totally enclosed and do not require ventilation openings in the housing for cooling. Instead the busway cools by radiation from the housing surface and by convection currents outside the housing. This method of cooling offers several advantages.

The totally enclosed busway needs no derating for different mounting positions, because it cools as efficiently in one position as another. Ventilated busways maintain their maximum operating temperatures within allowable limits by utilizing convection currents of air which pass through the housing itself to carry off excess heat. When the busway is mounted so that the perforated housing allows these convection currents to pass freely through the housing and between the conductors, this method of cooling is fairly efficient. However, if the busway is mounted in any position other than this *preferred position*, the busbars themselves interfere with this free passage of cooling air, efficiency is decreased, and the operating temperature rises.

Under these conditions, ventilated busway must be derated to a substantially lower current carrying capability. Or, if derating is unacceptable, oversized busbars must be used to reduce overall heating to an acceptable level.

Where totally enclosed construction is used on busways that are normally ventilated, even more stringent derating is required.

Plug-in switches or circuit breakers should be side mounted for maximum utilization. The preferred mounting position of most ventilated busway requires the plug-in units to be mounted on the top and bottom of the run, making those on top hard to get at, and making those on the bottom protrude into available headroom. Totally enclosed busway plug-in units may be side mounted for maximum utilization, without derating the busway.

The NE Code (Paragraph 364-6) requires that busways extend "vertically through dry floors if totally enclosed (unventilated), where passing through and for a minimum distance of 6 feet above the floor to provide adequate protection from physical damage."

Totally enclosed busway complies with this NE Code requirement with no modification. In the case of ventilated busway, if the enclosure is not provided by a busway manufacturer, this busway may meet requirements of NE Code Paragraph 364-6, but void the UL manifest, since UL cannot sanction modifications made to a product in the field. Ventilated busway requires expensive modification to satisfy both UL and NE Code requirements.

The safety precaution embodied in the NE Code requirement mentioned above is obvious. Totally enclosed construction affords much greater protection from mechanical damage to the busbars and insulation. It also gives much better protection from dust and dirt accumulation in the housing than does ventilated duct.

Because totally enclosed busway does not require that busbars be spaced apart for air flow between them, the physical size of the housing can, with proper design, be smaller, rating for rating.

The close spacing of busbars in this type of busway gives it exceptionally low reactance. This is particularly true of feeder busway, where the spacing between bars is less than 1/16 in. The low reactance thus achieved helps reduce voltage dips during the instant of a change in load, such as motor starting. Under such conditions, the high inrush current (up to 600% of load) is at a very low power factor and, consequently, the reactive component of voltage drop assumes greater importance.

If a busway can be designed to operate within satisfactory temperature rise limitations without relying on air convection currents through housing perforations, all the above advantages are gained. Ventilation is something one puts up with only when satisfactory operation cannot be achieved through more sophisticated design considerations.

Older designs of busway used several current-carrying nuts and bolts to connect each individual busbar. This was a natural consequence of the open busbar system out of which prefabricated busways grew. One of the most noticeable changes in recent designs is the elimination of this older joint in favor of an interleaved joint between sections, commonly called the one-bolt joint.

This type of joint incorporates one high-strength bolt to provide a clamping pressure of over 4000 lb. to the interleaved busbars. This force is distributed over the contact area by a 3-in. spring steel cup washer. Where two or three parallel conducting paths are provided, two or three similar bolts are used at each joint; but this arrangement is still referred to, in most industry literature, as a "one-bolt" type joint.

Most busway is UL listed for hanging on 10 ft. centers, eliminating half the hangers required by some older types. Most are also UL listed for hanging in any position including vertical risers.

SHORT-CIRCUIT BRACING

During a surge of current resulting from a low-resistance fault either in the busway or in the equipment fed by the busway, the conductors carrying the fault current are subjected to extremely large physical forces. These forces are the result of the interaction of the lines of magnetic flux that surround any current flow. For currents of the range that might be encountered during a bolted fault on a large busway system, these forces may reach values of several tons per lineal foot of conductor. For a three-phase system, there is always one conductor that is being forced away from the other two by these fault current forces, just as two magnets are repelled by each other when poles of like polarity are adjacent. To prevent physical damage to the busway, some means of restraining these forces must be provided.

Some busway is built with epoxilated construction to provide this restraint. After each bar is individually insulated, the sandwich of three or four busbars is wrapped together with a fiberglass tape, which is saturated with epoxy resin. The wrapped insulated bars are assembled in the steel and aluminum housing while the epoxy is still wet. As the epoxy sets, it not only bonds the conductor sandwich to the housing, which insures good heat transfer, but it also sets the glass tape into a rigid, strong encapsulation, confining short-circuit forces over the entire length of the busbars.

In a plug-in busway rated 800 A or higher, continuous short-circuit bracing is provided by clamping the edge of each 1/4-in. thick busbar in a groove or corrugation in the housing. Additional support is provided at the most critical area by building the molded insulator in such a way that it also braces the busbars against the forces of a fault current.

The support provided at the plug-on area is very important. Without it, movement of the busbars may damage plug-in units installed on the busway, even though the busway itself is able to withstand the fault and shows no apparent damage.

Lower ratings plug-in busway (225 A–600 A) do not need the corrugated housing (for heat transfer or bracing) required in the higher amperage. But they do use the same support insulator at the plug-in area, to provide bracing at this all-important point.

INSULATION

Most primary insulation of busway is Class B material, capable of satisfactory operation at temperatures up to 130°C. In feeder busway, a combination of Mylar polyester film and varnish-impregnated glass cloth is used. In plug-in busway, double layers of Mylar are used in the lower ratings. Higher ratings have an additional strip of mica tape around the edges of the busbars (where they are held in the housing corrugations). The use of all Class B insulation is intended primarily as a means of extending insulation life under normal operating conditions.

Nearly all insulating materials age more quickly when the operating temperature at which they are used approaches the allowed maximum.

Busbars on both plug-in and feeder busway are insulated over the entire length of the bar. This is necessary to prevent the propagation of traveling arcs. Should there be, by some mischance, an arc formed between busbars in busway, it would be confined to one length and would not damage adjacent lengths. Furthermore, the fault would be a low impedance path, because of the ionized air created at the arc. This would allow the overcurrent device protecting the run to operate rapidly, clearing the fault. Traveling arcs, common to noninsulated busbars, create extensive damage, because they run down the length of a busway and are a high impedance path for the fault current.

PLUG-IN BUSWAY

The typical plug-in busway provides a tap off every 2 ft. along both sides of the length—10 openings per 10-ft. length. Plug-in opening spacing is not disturbed by joint location on lengths of even footage. All openings are usable. Hanger location need not be considered when planning plug-in location or joint location. The one-bolt joint, continuous insulation, and plug-in opening insulator have already been discussed. This plug-in opening insulator, in addition to its function as a support for the busbars, also isolates each phase jaw from the others where a plug-in unit is attached. This effectively prevents accidental phase-to-phase or phase-to-housing shorts. The insulator and plug-in openings are covered by a hinged door held closed with a spring catch. By use of a screwdriver, this door may be opened prior to inserting the plug-in device so that the condition of the plug-in opening can be inspected. The hinged door has proven less susceptible to jamming and being painted shut than sliding or pivoting covers. Also, being able to open it prior to positioning the plug-in unit is a definite safety feature. A fouled opening would be detected before the plug-in unit was in place, not after it was partially installed. Nothing enters the busway housing but the plug-in jaws. No stab or probe is required. There is no chance of some steel part coming in contact with live parts. The conductors are set well back from the surface of the molded insulator to prevent accidental contact with them.

Plug-in busway ratings 800 A and larger employ a swing-away base feature, which allows bolt-on style units up to 1600 A capacity to be connected at any plug-in opening. Instead of opening the plug-in door, two screws are removed and the entire door assembly and plug-in opening insulator swing away, exposing a much wider busbar contact area.

PLUG-IN UNITS

All plug-in units are provided with a saw tooth grounding spring, which makes a positive ground connection between the plug body and the busway housing prior to jaw contact with the busbars. To complete the installation on most units, it is only necessary to tighten a single clamping screw, which rigidly fastens the plug enclosure to the busway housing. Tightening this clamping screw also releases the interlock so that the switching mechanism may be operated.

FUSIBLE UNITS

Fusible units through 400 A use the heavy-duty safety switch mechanism. This switch mechanism is quick-make, quick-break, independent of the operating handle, and incorporates visible blades, plated copper parts, arc suppressors, and a one-piece cross bar.

This mechanism operates in such a way that it is not possible to restrain the main contacts once the operating handle has started the closing action. The switch has positive action and may be opened or closed even if the main operating spring should be broken. All phase jaws are operated by the same solid one-piece cross bar.

When the cover of the plug-in unit is open, the position and condition of the switch blades can be seen. There is no question as to whether the switch is On or Off. Units may be positively padlocked in the Off position.

No live parts are exposed when the switch is off. The molded arc chamber barrier completely covers the line-side terminals. Heavy duty switches equipped with UL Class RK9 fuses have been tested satisfactorily on systems capable of delivering up to 100,000 symmetrical RMS amperes short-circuit current. All switches withstood the tests without any signs of failure.

The operating handle is mounted on the end of the plug-in box, not on the cover, and is always in control of the switch. Plug-in units are interlocked with the busway so that they must be turned off before being installed or removed. All plug-in units have interlocked doors. The door interlock can be overridden by use of a screwdriver so that the door can be opened while the switch is in the On position. With the switch door open and the clamping screw run into its normal clamped-to-busway position, the switch mechanism can be operated freely without interlocks of any kind. This allows maintenance personnel to operate the switch with the door open, without the necessity of using a free hand to operate a second interlock. Since such operation will normally be done while on a ladder, the necessity of using two hands to check switch operation would create a potentially dangerous situation.

Note that to remove the switch from the busway, the clamping screw must be backed out, which locks the interlock mechanism so that the switch cannot be turned on while disconnected from the busway.

All parts of the interlock mechanism are external and require no stab or probe to enter the busway housing. Since the interlock functions to prevent certain operations, the cause of interference and method of interlocking can be easily seen without opening the plug-in unit.

Switches in 400- and 600-A ratings can be ordered as cable-to-bus units for use as main disconnects.

CIRCUIT-BREAKER UNITS

Circuit-breaker plug-in units use molded case breakers in frame sizes from 100 A through 1600 A and in either standard- or high-interrupting capacity circuit breakers. Both standard- and high-interrupting type breakers are in the same size enclosure.

The interlocking mechanism is mounted on the end of the box as is done on the fusible units, and provides visual tripped indication as well as On, Off, and Reset positions. All line-side live parts are protected with a transparent polycarbonate shield. Auxiliary contacts are available on breaker units on special order. Breaker plug-in units in frame sizes of 225, 400, 1000, and 1600 A can be ordered for use as lug-to-bus main disconnect devices.

All switch and breaker plugs are equipped with an operator suitable for chain-pull or hook-stick operation from the floor.

WEATHERPROOF FEEDER BUSWAY

Feeder busway is manufactured in weatherproof as well as indoor construction. The weatherproof design incorporates gasketed covers for joint parts, vapor barriers, and other features that make it possible to install busway in exposed locations. Weatherproof busway can be connected to indoor feeder or to plug-in busway with the standard joint. Plug-in busway is manufactured only in an indoor configuration and should not be used outdoors.

No feeder or plug-in busway is suitable for extremely dusty or hazardous locations, or in extremely corrosive atmospheres. However, because of its totally enclosed housing, full insulation, and corrosive resistant tin plate, weatherproof busway is often more suitable for borderline cases than other types of busway.

TROLLEY BUSWAY

A number of types and ratings of plug-in and trolley busways are available for use as branch circuits or light-duty subfeeders. These include two-, three-, and four-pole types for supplying such loads as lighting fixtures, portable electric tools, cranes and hoists. Such busways have a continuous opening for insertion of plugs or trolley conductors, which can readily move back and forth on busbars.

Where a busway is used as a feeder, devices, or plug-in connections for tapping off subfeeder or branch circuits from the busway contains the overcurrent devices required for the protection of the subfeeder of branch circuits. The plug-in device consists of an externally operable circuit breaker or an externally operable fusible switch. Where such devices are mounted out of reach and contain disconnecting means, suitable means such as ropes, chains, or sticks are provided for operating the disconnecting means from the floor.

A typical system of branch circuit plug-in busway is rated at 50 A and provides for plug-connection of incandescent, fluorescent or HID lighting, or both, at any point along the length of the busway. Fusible plugs or circuit breaker plugs can be inserted at any point to feed cord-connected loads.

Another continuous-connection branch-circuit busway system is made with two different ratings—20 A or 50 A. This system provides for either plug-in connection or moving trolley connection to the busbars within the assembly. As with other branch-busway systems, this system offers great flexibility in supplying light loads such as lighting or small machines. In commercial areas, schools and other labs, workshops and a wide variety of industrial locations, lighting fixtures can be moved, added or removed in a matter of minutes, without rewiring, using extra materials, or losing time. This is particularly advantageous for areas where the lighting must be altered since work areas are regularly changed or shifted.

With all of the branch-busway systems, a complete line of accessories are made to fully use the busway. Typical accessories include: feed-in boxes for supplying power to the busway, couplings, elbows to provide for change in direction of a busway run, end caps to close the ends of busway, a variety of plug-in or trolley connectors, or both, for tapping power from the busway, plug-in outlet boxes, plug-in receptacle outlets, hangers for supporting the busway in the installed position and hardware for supporting lighting fixtures from the busway.

SPECIAL BUSWAYS

In addition to the standard busway equipment available, special busways can be provided. For instance, although many standard busways can be used on high-frequency circuits (that is, for frequencies above 60 cycles per second), either with or without derating of the allowable current value permitted for 60 cycles per second, special high-frequency busway (either feeder or plug-in type) can be provided for frequencies up to many thousands of cycles per second. Or, for high-capacity direct-current circuits rated to more than 10,000 A, busway can be used.

APPLICATION

Although busway has wide application potential, it is limited to use for *exposed*—not encased in a housing—work only. Some exception to this rule is frequently made where busway is not permanently concealed by building finish, which cannot be readily removed. Use of busway above suspended ceilings of tiles that can be easily removed to provide access to the busway is sometimes permitted.

Busway must not be used where exposed to severe physical damage or corrosive vapors. Special finishes for busway are made to permit use in mildly corrosive atmospheres, but not in highly corrosive atmospheres or in battery rooms. Busway must not be embedded in the ground or in concrete, and it must not be used in hazardous locations where explosive gases or vapors, or ignitable dusts or fibers are present. Busway must not be used outdoors or in damp or wet locations unless specially approved for the purpose. Busway may be used as the service entrance conductors for a building.

Ambient temperature must be considered wherever busway is installed. The construction and current rating of busway is based on a certain maximum ambient temperature, such as 40°C. When used in higher ambients than what it is rated for, busway must be carefully derated to carry less current in accordance with manufacturer's instructions.

In the installed position, busways must be supported at intervals of not over 5 ft, unless the busway is specifically approved for use with supports at greater intervals. In no case, however, may the spacing of supports exceed 10 ft. When a busway is supported in a vertical position, such as in an office building riser, the supports for the busway must be designed for vertical installation.

Busway runs must be protected against overcurrent by protective devices installed at the supply side. When a reduced size of busway is tapped from a larger busway, the general rule is that overcurrent protection must be installed at the tap point to protect the smaller size of busway if the overcurrent device protecting the larger busway does not protect the smaller busway. There is one notable exception, however. If the reduced size of busway tapped from a larger busway does not extend more than 50 ft. and has a rating at least one third the rating or setting of the overcurrent device protecting the larger busway, the reduced size of busway does not have to have overcurrent protection at its point of tap from the larger busway.

12

Transmission Systems and Substations

Alternating-current transmission and distribution systems are not normally restricted by distance conditions. It is economically feasible to transmit almost any value of power over any reasonable distance with alternating current. This is primarily because the voltage may be transformed—either up or down—with stationary transformers, which are relatively inexpensive and are very efficient in operation. Furthermore, transformers have no moving parts, so maintenance is kept to the minimum. Therefore, in application, electrical energy may be generated at the most convenient voltage, increased with a step-up transformer to a voltage high enough for economical transmission or distribution and then lowered at the distant end of the line with a step-down transformer to a voltage suitable for effective utilization.

In general, a four-wire, three-phase transmission system is more economical than a single-phase system. Therefore, for long-distance systems, the three-phase transmission is almost always adopted.

Many different voltages have been used on the ac systems in the United States and Canada, but in the past several decades, standard voltages have been adopted in most areas to decrease the cost of transformers, generators, and other equipment.

The source of most commercial electric energy is a generator or a combination of generators or alternators. The generator is driven either by engines, hydraulics, or steam, although wind power is again being experimented with. Steam, of course, can be produced by coal, oil, or gas fired generators, or nuclear fission.

Most generating stations utilize an outdoor substation, which contains step-up transformers to transform the generator voltage of say, 14 kV to the transmission voltage, which may be as high as 750 kV. A three-phase transformer is normally used for each generator.

Other equipment contained in the outdoor substation includes disconnect switches for isolating circuit breakers, line-grounding disconnect switches, and lightning arresters. Current and potential instrument transformers are also usually located in the outdoor substation to supply operating power for protective relays and metering devices.

The main transmission lines between the generating station and the bulk power station are usually protected by differential protective relays, which measure the incoming and outgoing currents. The relays operate to open the line circuit breakers, usually placed at each end of the line, when a fault occurs and the incoming and outgoing currents do not balance. This way, any faults may be quickly isolated to maintain system stability and preserve service to the unfaulted portion of the system.

The outdoor bulk power substation is very similar to the generating station substation in that it contains circuit breakers, the outdoor substation, and other related switchgear equipment. At the generating station substation, the generated voltage is stepped up to the transmission voltage. However at the bulk power substation, this voltage must again be stepped down, or reduced, by the use of step-down transformers. From this point, power is supplied to distribution substations, and each of these subtransmission circuits are protected by some form of overcurrent protection.

Substation and primary feeder circuits have many variations and voltages. The actual voltage used is governed by a large number of factors, which are mainly economic but also include such practical matters as the area to be served, load densities, estimated future growth, terrain, and availability of rights-of-way and substation sites.

Various designs of utilization substations are also needed to provide the many types of services that are in demand. Typical loads include residential services, which are usually 240/120 V, single phase; and commercial services, which may be 240/120 V, single phase, or perhaps 208/120 Y, three-phase services; and many industrial sites will require, say, 13 kV.

The local utility companies may require synchronous condenser or capacitor substations to maintain better system voltage, or some feeders may supply a low-voltage network where network protectors are used. Such circuits are usually equipped with some form of fault detector and with an automatic-reclosing switching scheme for automatic restoration of service after an outage.

Conversion substations are used in some areas to change alternating current to direct current for services such as 600, 1500, or 3000 V for use by railways such as the Pennsylvania Railway. Certain types of mills and electrochemical plants also require direct current for their manufacturing processes. Sometimes the plant itself provides this conversion equipment while the local power companies provide the conversion in other cases.

PROTECTION OF POWER SYSTEMS

A variety of situations may interfere with the normal operation of a power system. The predominant abnormal conditions on distribution circuits are line faults, system overloads, and equipment failures. Atmospheric disturbances and both animal and human interference with the system are generally the underlying causes of these conditions.

Line faults can be caused by strong winds, which whip phase conductors together or which blow tree branches on the lines. In winter, freezing rain can produce a gradual buildup of ice on a circuit, and eventually one or more conductors may break and fall to the ground. Squirrels and other animals sometimes place themselves between an energized portion of the circuit and ground, which can fault the system.

On underground systems, cables severed by earth-moving equipment are a prevalent cause of faults. Lightning strokes can fault a system by opening lines or initiating arcs between conductors. Unforeseen load growth is the primary cause of overloads. Equipment failure can be caused by lightning; insulation deterioration; improper design, manufacture, installation, or application; and system faults.

SECONDARY DISTRIBUTION SYSTEMS

Secondary distribution systems are installed and protected in a similar fashion to the higher-voltage primary distribution systems. The more important lines in the secondary system will require differential protection, while the less important ones may need nothing more than conventional overcurrent protection. However, the same rules apply: A fault must be quickly removed from the system by tripping the minimum number of circuit breakers or blowing the minimum number of fuses, leaving the balance of the system in operation.

There are two general arrangements of transformers and secondaries used. The first arrangement is the sectional form, in which a unit of load, such as one city street or city block, is served by a fixed length of secondary, with the transformer located in the middle. The second arrangement is the continuous run in which transformers are spaced along the load at the most suitable points. As the load grows or shifts, the transformers spaced along it can be moved or rearranged, if desired. In sectional arrangement, such a load can be cared for only by changing to a larger size of transformer or installing an additional unit in the same section.

One of the greatest advantages of the secondary bank is that the starting currents of motors are divided between transformers, which reduces voltage drops and also diminishes the resulting lamp flicker at the various outlets.

In the sectional arrangement, each transformer feeds a section of the secondary and is separate from any other. If a transformer becomes overloaded, it is not helped by adjacent transformers; rather, each transformer acts as a unit by itself. Therefore, if a transformer fails, there is an interruption in the distribution service of the section of the secondary distribution system that it feeds. This is the layout used most frequently for secondary distribution systems at the present time.

Power companies all over the United States are now trying to incorporate networks into their secondary power systems, especially in areas where a high degree of service reliability is necessary. Around cities and industrial sites, most secondary circuits are three-phase—either 120/208 V, Y connected or 480/208 V, Y connected. Usually, two to four primary feeders are run into the area, and transformers are connected alternately to them. The feeders are interconnected in a grid, or network, so that if any feeder goes out of service, the load is still carried by the remaining feeders.

To protect a grid-type power system, a network protector is usually installed between the transformer and the secondary mains. This protector consists of a low-voltage circuit breaker controlled by relays, which cause it to open when reverse current flows from the secondaries into the transformer and to close again when normal conditions are restored. If a short circuit or ground fault should occur on a primary feeder or on any transformer connected to it, the feedback of current from the network into the fault through all the transformers on that feeder will cause the protective switches to open, disconnecting all ties between that feeder and the secondary mains. When the trouble is repaired and normal voltage conditions are restored on that feeder, the switches will reclose, putting the transformers back into service.

Designers of a network are always cautious about the placement of transformers. The transformers should be large enough and close enough together to be able to burn off a ground fault on the cable at any point. If not, such a fault might continue to burn for a long time.

The primary feeders supplying networks are run from substations at the usual primary voltage for the system, such as 4160, 4800, 6900, or 13,200 V. Higher voltages are practicable if the loads are large enough to warrant them.

Network power systems are usually installed with primaries, secondaries, and transformers all underground. The transformers and secondaries may, however, be overhead, or they may use a combination of overhead and underground construction.

SECONDARY SERVICES

The 4160-V transformer came about through the 4160-V connection on 2400-V transformers. In some cases, it was advantageous to connect transformers between phase wires on a 2400/4160Y-V system, and this required a transformer having a winding voltage of 4160 V. These 4160-V transformers are now used in several ways. First, they are used in three-phase delta banks connected to 2400/4160Y-V systems. Another application is on 4160-V single-phase lines taken off of a 2400/4160Y-V three-phase system, necessitating the use of 4160-V transformers.

In some instances, the 4160-V transformers are used for rural systems rated 4160/7200Y. With this system, 4160-V transformers can be used between phase wire and neutral of a three-phase, four-wire system, and 7200-V transformers can be used between phase wires.

The 4800-V transformers are frequently used in some sections of the United States where distribution circuits run through thickly populated rural and suburban areas. Distribution lines in these localities are necessarily much longer than in cities, and the 2400-V system doesn't have a high enough voltage to be economical. On the other hand, 4800-V distribution systems in these areas have proved to be quite logical and satisfactory.

Again, the systems originally were 4800-V delta, three-phase systems with 4800-V, single-phase branch lines. However, in many cases, these delta systems are now being converted to 4800/8320Y, giving a higher system voltage while still using the same equipment that was used on 4800-V delta systems.

Rural electrification in thinly populated areas requires still higher voltage for good performance and economy. Therefore, for rural power systems in certain sections of the United States, 7200-V distribution systems have been used quite extensively and successfully. The early rural systems were 7200 V delta, three phase in most cases, with 7200-V branch lines. These systems are now giving way to 7200/12,470Y-V, three-phase, four-wire systems. In fact, this is probably the most popular system in use today.

Although less popular than the 7200-V class of transformers, 7620-V systems are sometimes used for rural electrification. Most of these systems are actually 7620/13,200Y, three-phase, four-wire systems. On this type of system, 7620-V, single-phase transformers can be used between phase wire and neutral of the three-phase system, or 13,200-V transformers can be used between phase wires. This type of system works out very economically for power companies that have both 7620- and 13,200-V distribution systems. In this situation, transformers can be used on either system, thereby making stocks of transformers flexible.

There are some 12,000-V, three-phase delta systems that were installed some time ago for transmission and power over greater distances than were feasible in lower voltages. There are now two applications for 12,000-V transformers. The first is for use on 12,000-V delta systems and the second for use on 7200/12,470Y-V systems.

The 13,200-V transformers also have two applications. First, they can be used on distribution systems that are 13,200 V delta, three-phase, which were built to distribute electrical energy over considerable distances. The second application has already been mentioned in connection with the 7620/13,200-V, three-phase, four-wire system. On this system, the 13,200 standard transformer can be used between phase wire of the three-phase, four-wire system. This connection is made quite often when it is necessary to connect a three-phase bank of transformers to the 7620/13,200Y-V systems.

In addition to use in rural areas, the 12,000- and 13,200-V distribution systems are often used in urban areas. In relatively large cities having considerable industrial loads, 13,200- and 12,000-V lines are frequently run to serve industrial loads, while the 2400/4160Y-V system is used for the residential and commercial load.

Units that are 24,940-Grd-Y/14,000 V have one end of the high-voltage winding grounded to the tank wall and are suitable only for use on systems having the neutral grounded throughout its length.

System voltages to 68 kV have been designated as distribution, although transmission lines also operate at these same voltages. The trend is to convert these lines to four-wire distribution systems and use transformers with primary windings connected in wye. As an example, a multigrounded neutral might be added to a 34,500-V system, and 20,000-V transformers would be used to supply the customers. The system is also in use for new construction in high-load-density areas.

USE OF CAPACITORS

A capacitor is a device that will accept an electrical charge, store it, and release the charge when desired. In its simplest form, it consists of two metallic plates separated by an air gap. The larger the surface of the plates and the closer they are together, the greater the capacity will be.

If the space between the plates is filled with various insulating materials, such as kraft paper, linen paper, or oil, the capacity will be greater than with air. The increase in this capacity for any specific material as compared to air is called the dielectric constant and is expressed by the letter K. If plate area and spacing remain unchanged but a certain grade of paper gives a capacitor twice the capacity it had with air, then the value for K would be 2. All insulating materials have a value of K, which merely expresses its dielectric effectiveness as compared to air. Thus, only three factors govern capacity

1. How big are the plates?
2. What material separates them?
3. How close are the plates together?

Capacitors operate on both dc and ac circuits. If connected to the terminals of a dc circuit, a pair of plates without a charge on them will accept a static charge. Current will rush into the capacitor until each plate is at the potential of the line to which it is connected. Once this potential is reached (a very rapid process), no further current, other than leakage current, will flow. If removed from the line, the capacitor plates will maintain their charge until it is dissipated by leakage between plates or by deliberate contact between plate terminals. This principle is used in surge generators. In the ac application, the plates are alternately charged and discharged by the voltage changes of the circuit to which they are connected. It is this condition that makes possible the use of capacitors for power-factor correction.

POWER FACTOR

Power factor is the ratio of useful working current to total current in the line. Since power is the product of current and voltage, the power factor can also be described as a ratio of real power to apparent power and be expressed as

$$\text{Power Factor} = \text{kW}/\text{kVA}$$

Apparent power is made up of two components: (1) real power (expressed in kilowatts), and (2) the reactive component (expressed in kilovars). This relationship is shown in Fig. 12-1. The horizontal line AB represents the useful real power (kW) in the circuit. The line BC represents the reactive component (kvar) as drawn in a downward direction. Then a line from A to C represents apparent power (kilovolt-amperes or kVA). To the uninitiated, the use of lines representing quantity and direction is often confusing. Lines utilized as such are called vectors. Imagine yourself at point A desiring to reach point C,

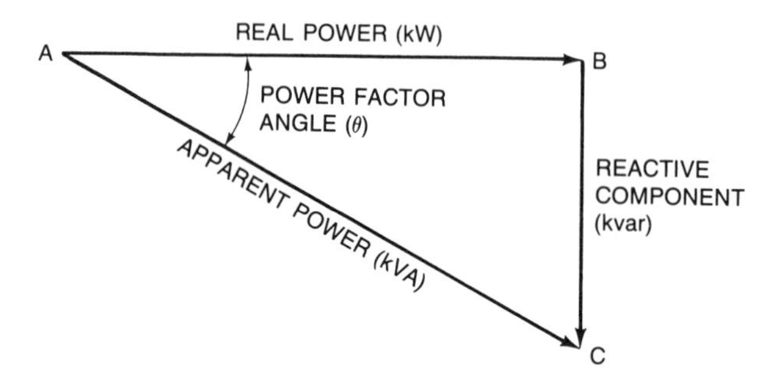

Figure 12-1 Relationship of real power and the reactive component.

but, due to obstructions, you must first walk to *B*, and then turn a right angle and walk to *C*. The energy you dissipated in reaching *C* was increased because you could not take the direct course, but, in the final analysis, you ended up at a point the direction and distance of which can be represented by the straight line *AC*. The power-factor angle shown is called theta (θ).

There is interest in power factor because of the peculiarity of certain ac electrical equipment requiring power lines to carry more current than is actually needed to do a specific job unless a principle that has long been understood but not until recent years given the attention it deserves is utilized. This principle utilizes the application of capacitors.

LEAD AND LAG

Real and reactive components cannot be added arithmetically. To understand why, consider the characteristics of electrical circuits. In a pure resistance circuit, the alternating voltage and current curves have the same shape, and the changes occur in perfect step, or phase, with each other. Both are at zero with maximum positive peaks and maximum negative peaks at identical instants. Compare this with a circuit having magnetic characteristics involving units such as induction motors, transformers, fluorescent lights, and welding machines. It is typical that the current needed to establish a magnetic field lags the voltage by 90°.

Visualize a tube of toothpaste. Pressure must be exerted on the tube before the contents oozes out. In other words, there is no flow until pressure is exerted, or, analogously, the flow (current) lags the pressure (voltage). In like manner, the magnetic part of a circuit resists, or opposes, the flow of current through it. In a magnetic circuit, the pressure precedes or leads the current flow, or, conversely, the current lags the voltage.

Peculiarly, in a capacitor, we have the exact opposite: Current leads the voltage by 90°. Visualize an empty tank to which a high-pressure air line is attached by means of a valve. At the instant the valve is opened, a tremendous rush of air enters the tank, gradually reducing in rate of flow as the tank pressure approaches the air line pressure. When the tank is up to full pressure, no further flow exists. Accordingly, you must first have a flow of air into the tank before it develops an internal pressure. Consider the tank to be a capacitor and the air line to be the electrical system. In like fashion, current rushes into the capacitor before it builds up a voltage, or, in a sense, the current leads the voltage.

A water system provides a better comparison. The system shown in Fig. 12-2 is connected to the inlet and outlet of a pump. For every gallon that enters the upper section, a gallon must flow from the lower section as the plunger is forced down. If the pressure on the upper half is removed, the stored energy in the lower spring will return the plunger to midposition. When the direction of flow is reversed, the plunger travels upward, and if flowmeters were connected to the inlet and outlet, the inlet and outlet flows would prove to be equal. Actually, there is no flow through the cylinder but merely a displacement. The only way we could get a flow through the cylinder would be by leakage around the plunger or if excessive pressure punctured a hole in the plunger.

In a capacitor, we have a similar set of conditions. The electrical insulation can be visualized as the plunger. As we apply a higher voltage to one capacitor plate than to its companion plate, current will rush in.

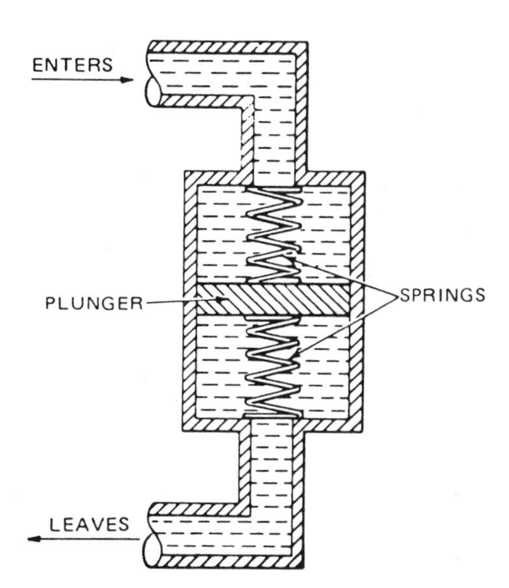

Figure 12-2 A water system used to illustrate lead and lag in an electrical system.

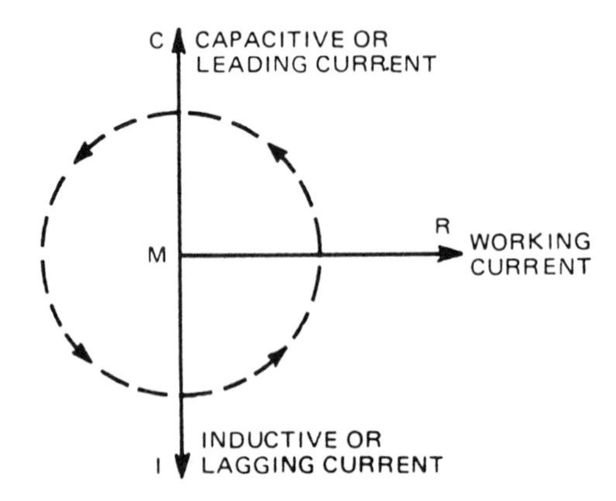

Figure 12-3 Diagram representing three different kinds of current: phase, lagging, and leading.

If we remove the pressure and provide an external conducting path, the stored energy will flow to the other plate, discharging the capacitor and bringing it to a balanced condition in much the same fashion as the spring returned the plunger to midposition.

Since no insulator is perfect, some leakage current will flow through it when a voltage differential exists. If we raise the voltage to a point where we break down or puncture the insulation, then the capacitor is damaged beyond repair and must be replaced.

Cognizance of the three different kinds of current discussed, namely, in phase, lagging, and leading, permits the drawing of the relationship in Fig. 12-3.

The industry accepts a counterclockwise rotation about point M as a means of determining the relative phase position of voltage and current vectors. These may be considered as hands of a clock running in reverse. MC is preceding MR; hence, it is considered leading. MI follows MR; therefore, it is lagging.

The two angles shown are right angles (90°); therefore, it becomes apparent that MC and MI are exactly opposite in direction and will cancel out each other if of equal value.

Consider a wagon to which three horses (C, I, R) are hitched, as in Fig. 12-4. If C and I pull with equal force, they merely cancel one another's effort, and the wagon will proceed in the direction of R, the working horse. If only R and I are hitched, the course will lie between these two.

Most utility lines contain quantities of working current and lagging current, and one type is just as effective in loading up the line as the other. If we know how much inductive current a line is carrying, then we can connect enough capacitors to that line to cancel out this wasteful and undesired component.

Just as a wattmeter will register the kilowatts in line, a varmeter will register the kvar of reactive power in the line. If an inductive cir-

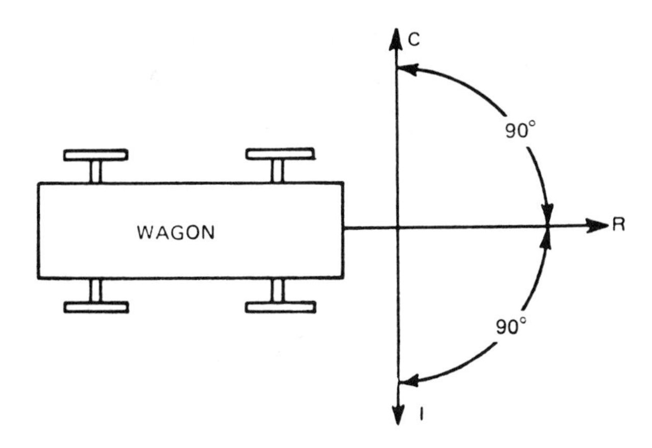

Figure 12-4 Horse and wagon illustrating phase, lagging, and leading current.

cuit is checked by a meter that reads 150,000 var (150 kvar), then application of a 150-kvar capacitor would completely cancel out the inductive component, leaving only working current in the line.

POWER-FACTOR CORRECTION

In actual practice, full correction to establish the unity power factor is rarely, if ever, recommended. If a system had a constant 24-hr load at a given factor, such correction could be readily approached. Unfortunately, such is not the case, and we are faced with peaks and valleys in the load curves.

If we canceled out, by the addition of capacitors, the inductive (lagging) kvars at peak conditions, our capacitors would continuously pump their full value of leading kilovars into the system. Thus, during early morning hours when inductive kvars are much below peak conditions, a surplus of capacitive kvars would be supplied, and a leading power factor would result. Local conditions may justify such overcorrection, but, in general, overcorrection is not recommended. Figure 12-5 will show how far we can go with fixed capacitors.

A recording kilovarmeter can readily give the curve shown, or else it can be calculated if we know the kW curve and the power factor throughout the day. From 2 to 6 a.m., refrigerators, transformer excitation, high factory loads, and the like result in relatively low readings. When the community comes to life in the morning, televisions, radios, appliances, and factory loads build up a high inductive kvar peak.

The area under the shaded section, which is limited by the lowest 24-hr kvar reading, represents the maximum degree of correction to

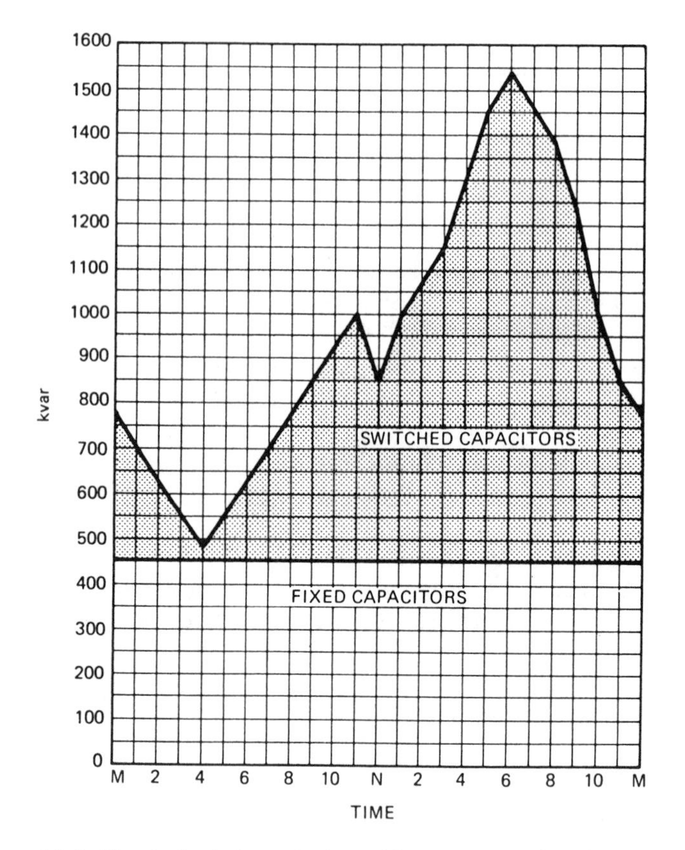

Figure 12-5 Graph depicting relation of kvar to time with fixed capacitors.

which fixed capacitors are generally applied. It becomes apparent that if we go beyond this point, the utility would be faced with leading power-factor conditions during light-load periods. The shaded area falls into the zone that can be handled effectively with switched capacitors.

SWITCHED CAPACITORS

There are several ways in which switching of capacitors can be accomplished. A large factory would arrange by manual or automatic operation to switch in a bank of capacitors at the start of the working day and disconnect them when the plant shuts down. The energizing of a circuit breaker control coil can be effected with var, current, voltage, temperature, and time controls.

JUSTIFICATION FOR CAPACITORS

The application of capacitors to electrical distribution systems has been justified by the overall economy provided. Loads are supplied at reduced cost. The original loads on the first distribution systems were predominantly lighting so the power factor was high. Over the years, the character of loads has changed. Today, loads are much larger and consist of many motor-operated devices that place greater kilovar demands upon electrical systems. Because of the kilovar demand, system power factors have been lower.

The result may be threefold

1. Substation and transformer equipment may be taxed to full thermal capacity or overburdened.
2. High kilovar demands may, in many cases, cause excessive voltage drops.
3. A low power factor may cause an unnecessary increase in system losses.

Capacitors can alleviate these conditions by reducing the kilovar demand from the point of demand all the way back to the generators. Depending on the uncorrected power factor of the system, the installation of capacitors can increase generator and substation capability for additional load at least 30% and can increase individual circuit capability, from the standpoint of voltage regulation, 30% to 100%.

13

Transformers

Electric power produced by alternators in a generating station is transmitted to locations where it is utilized and distributed to users. Many different types of transformers play an important role in the distribution of electricity. Power transformers are located at generating stations to step up the voltage for more economical transmission. Substations with additional power transformers and distribution transformers are used to step down the voltage to a level suitable for utilization.

Two coils or windings on a single magnetic core form a transformer. Such an arrangement will allow transforming a large alternating current at low voltage into a small alternating current at high voltage or vice versa. Transformers, therefore, make it possible to change from a generator that produces moderately large alternating currents at moderately high voltages to one that produces very high voltage and proportionately small current in the transmission lines, which permits the use of smaller cable and allows less power loss.

The essential parts of a transformer are a laminated iron core upon which are wound two separate insulated coils—the primary and the secondary as shown in Fig. 13-1. In most cases, the primary coil is connected to the supply or main side of the line where the alternating current sets up an alternating magnetic flux. The action not only sets up a countervoltage equal and opposite in the primary coil but also sets up a voltage in the secondary coil. The ratio of the voltage in the

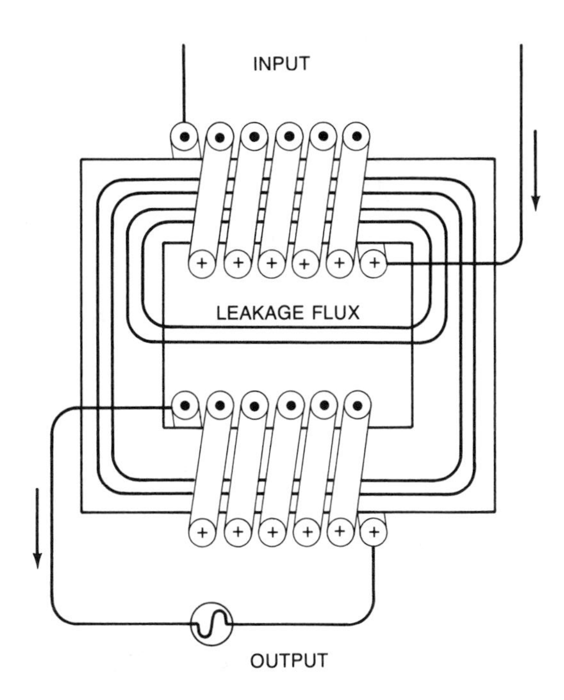

Figure 13-1 Essential parts of a transformer.

secondary coil as compared to that in the primary coil depends on the amount of magnetic flux, the frequency of the alternating current, and, mainly, the number of turns in the coils.

In a well-designed transformer, there is very little magnetic leakage. The effect of the leakage is to cause a decrease of secondary voltage when the transformer is loaded. When a current flows through the secondary in phase with the secondary voltage, a corresponding current flows through the primary in addition to the magnetizing current previously mentioned. The magnetizing effects of the two currents are equal and opposite.

In a perfect transformer—one without eddy-current losses, resistance in its windings, and magnetic leakage—the magnetizing effects of the primary load current and the secondary current neutralize each other, leaving only the constant primary magnetizing current effective in setting up the constant flux. If supplied with a constant primary pressure, such a transformer would maintain constant secondary pressure at all loads. Obviously, the perfect transformer has yet to be built; the closest is a transformer with very small eddy-current losses where the drop in pressure in the secondary windings is not more than 1% to 3% depending on the size of the transformer.

INSULATING TRANSFORMERS

Insulating transformers are those with no internal electrical connection between the primary and the secondary windings. In all but a few designs, it is customary first to wind the low-voltage coil next to the core and, once completed, finish the unit by winding the high-voltage coil over it. This construction places the conductors energized at the high voltage a greater physical distance from the magnetic core, which is normally grounded. The core is electrically interconnected with core clamps, steel structure, and enclosing case, all of which are connected with a ground lead to the plant or system ground.

In the usual high-voltage transformer design, the lower voltage coil is wound next to the core. This in turn is covered with mica sheet layer insulation over which the high-voltage coil is wound. Additional sheet mica insulation is applied around each coil with a final wrap of glass tape for extra electrical and mechanical strength. This is commonly known as barrel-type construction with a 220°C insulation system.

Cooling ducts are strategically placed within each winding to carry away the internally generated heat. The smooth exterior coil surfaces in a vertical plane minimize the accumulation of dirt.

TAPS

If an electric utility could always guarantee to deliver exactly the rated primary voltage at every transformer location, taps would be unnecessary. However, it is not possible to achieve this, and in recognition of this fact, the public service commissions of each state allow reasonable variations above or below the nominal value.

Generally speaking, if a load is very close to a substation or generating plant, the voltage will consistently be above normal. Near the end of the line the voltage may be below normal. The primary taps are used to match these voltages.

In large transformers, it would be very inconvenient to move the thick, well-insulated primary leads to different tap positions when changes in source-voltage levels make this desirable. Therefore, taps such as those shown in the wiring diagram in Fig. 13-2 are used. In this transformer, the permanent high-voltage leads would be connected to H_1 and H_2 and the secondary leads, in their normal fashion, to X_1, X_2, X_3, and X_4. Note, however, the tap arrangement available at 2, 3, 4, 5, 6, and 7. Until a pair of these taps is interconnected with a jumper, the primary circuit is not completed. If this were the typical 7200Y primary, for example, the transformer would have a normal 1620 turns.

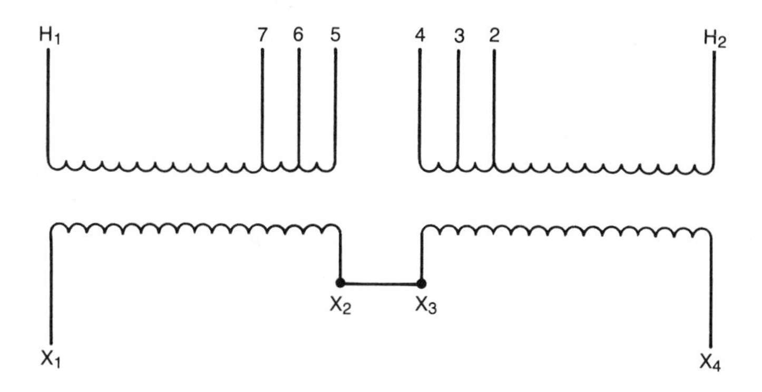

Figure 13-2 Wiring diagram of a transformer.

Assume 810 of these turns are between H_1 and 6 and another 810 between 3 and H_2. Then if we connect 6 and 3 together with a flexible jumper on which lugs have already been installed, the primary circuit is completed, and we have a normal ratio transformer that could deliver 120/240 V from the secondary.

Between tap 6 and either 5 or 7, we have the familiar 40 turns. From previous examples, changing the jumper from 3 and 6, to 3 and 7 removes 40 turns from the left half of the primary. The same condition would apply on the right half of the winding if the jumper were between 6 and 2. Either connection would boost secondary voltage by 2-1/2%. Had we connected 2 and 7, 80 turns would have been omitted and a 5% boost would have resulted. Placing the jumper between 6 and 4, or 3 and 5 would reduce output voltage by 5%.

PARALLEL OPERATION OF TRANSFORMERS

Three-phase transformers, or banks of single-phase transformers, may be connected in parallel provided each of the three primary leads in one three-phase transformer is connected in parallel with a corresponding primary lead of the other transformer. The corresponding leads are the leads that have the same potential at all times and the same polarity. Furthermore, the transformers must have the same voltage ratio and the same impedance voltage drop.

When three-phase transformer banks operate in parallel and the three units in each bank are similar, the division of the load can be determined by the same method previously described for single-phase transformers connected in parallel on a single-phase system.

In addition to the requirements of polarity, ratio, and impedance, paralleling of three-phase transformers also requires that the angular displacement between the voltages in the windings be taken into consideration when they are connected together.

Phasor diagrams of three-phase transformers that are to be paralleled greatly simplify matters. With these, all that is required is to compare the two diagrams to make sure they consist of phasors that can be made to coincide; then connect together terminals corresponding to coinciding voltage phasors. If the diagram phasors can be made to coincide, leads that are connected together will have the same potential at all times. This is one of the fundamental requirements for paralleling.

AUTOTRANSFORMERS

The basic difference of an autotransformer and a double-wound transformer is that an autotransformer is a transformer whose primary and secondary circuits have part of a winding in common and therefore the two circuits are not isolated from each other. The application of an autotransformer is a good choice for some users where a 480Y/277- or 208Y/120-V, three-phase, four-wire distribution system is utilized. The main advantages are as follows:

1. Lower purchase price
2. Lower operating cost due to lower losses
3. Smaller size, easier to install
4. Better voltage regulation
5. Lower sound levels

An autotransformer, however, cannot be used on a 480- or 240-V, three-phase, three-wire delta system. A grounded neutral phase conductor must be available in accordance with NE Code Article 210-9.

CONTROL TRANSFORMERS

The term control transformers could be used to describe a large variety of transformers for many different purposes. However, for our use, control transformer will be defined as a device used to reduce supply voltages to 120 V or lower for the operation of electromagnetic devices such as contactors, solenoids, and relays.

Industrial control transformers are especially designed to accommodate the momentary current inrush caused when electromagnetic components are energized—without sacrificing secondary voltage stability beyond practical limits.

Most control transformers are dry-type step-down units with the secondary control circuit isolated from the primary line circuit to assure maximum safety. These transformers and other components are usually mounted within an enclosed control box or control panel, which has push-button stations independently grounded as recommended by the NE Code and other safety codes.

Other types of control transformers—sometimes referred to as control and signal transformers—normally do not have the required industrial transformer regulation characteristics. Rather, they are constant-potential, self-air-cooled transformers used for the purpose of supplying the proper voltage (usually reduced) for control circuits of electrically operated switches or other equipment and, of course, for signal circuits. Some are of the open type with no protective casing over the windings, while others are enclosed and have a metal casing over the winding.

Potential Transformer

In general, a potential transformer supplies low voltage to an instrument that is connected to its secondary. The voltage is proportional to the primary voltage, but it is small enough to be safe for the test instruments. The secondary of a potential transformer may be designed for several different voltages, but most are designed for 120 V.

As may be imagined, these transformers are used with devices requiring voltage for operation, such as voltmeters, frequency meters, power-factor meters, and watt-hour meters. In the case of a multimeter, one transformer may be used for any number of instruments at the same time, provided the total current usage does not exceed the rating of the potential transformer.

The potential transformer is primarily a distribution transformer especially designed for good voltage regulation, so that the secondary voltage under all conditions will be as nearly as possible a definite percentage of the primary voltage.

Current Transformer

A current transformer is an instrument transformer that is normally used to supply current or voltage of a smaller value than the line current or voltage to an electrical instrument.

A current transformer supplies current to an instrument connected to its secondary, the current being proportional to the primary current but small enough to be safe for the instrument. The secondary of a current transformer is usually designed for a rated current of 5 A.

A current transformer operates in the same way as any other transformer; that is, the same relation exists between the primary and the secondary current and voltages. A current transformer is connected in series with the power lines to which it is applied, so that line current is connected to current devices such as ammeters, wattmeters, watt-hour meters, power-factor meters, some forms of relays, and trip coils of some types of circuit breakers.

When no instruments or other devices are connected to the secondary of the current transformer, a short-circuit device or connection is placed across the secondary. In other words, the secondary circuit of a current transformer should never be opened while the primary is carrying current. Before disconnecting an instrument, the secondary of the current transformer must be short circuited. If the secondary circuit is opened while the primary winding is carrying current, there will be no secondary ampere turns to balance the primary ampere turns, so the total primary current becomes exciting current and magnetizes the primary and secondary windings. Since, to secure accuracy, current transformers are designed with normal exciting currents of only a small percentage of full-load current, the voltage produced with the secondary open circuited is high enough to endanger the life of anyone coming in contact with the meter or leads. Also, the high secondary voltage may overstress the secondary insulation and cause a breakdown. Operation with the secondary open circuited may also cause the transformer core to become permanently magnetized. If this should occur, the core may be demagnetized by passing about 50% excess current through the primary, with the secondary connected to an adjustable high resistance that is gradually reduced to zero.

SATURABLE-CORE REACTORS

A saturable-core reactor is a magnetic device that has a laminated iron core and ac coils similar in construction to a conventional transformer; it is uniquely effective in the control of all types of high power-factor loads. The coils in a saturable-core reactor are called gate windings. In addition, it is designed with an independent winding by which direct current is introduced for control.

When the ac coils of a saturable-core reactor are carrying current to the load, an ac flux (magnetism) saturates the iron core. With only ac coils functioning, the magnetism going into the iron core restricts the ac flow voltage output to the load to about 10% of the line supply voltage. Since the iron core is always fully saturated with magnetic flux, the use of direct current from the control winding introduces dc flux, which displaces the ax flux from the iron core. This action reduces the impedance and causes the voltage output to the load to increase. By adjusting the dc flux saturation, the impedance of the ac gate winding may be infinitely varied. This provides a smooth control ranging from approximately 10% to 94% of the line voltage at the load. Since there is practically no power loss in the control of the impedance in a saturable-core reactor, a relatively small amount of direct current can control large amounts of alternating current.

Saturable-core reactors eliminate the need for mechanical and resistance controls and are a very efficient means of proportional power control for resistance heating devices, vacuum furnaces, infrared ovens, process heaters, and other current-limiting applications. In lighting control, especially where wattage per circuit is large, a saturable-core reactor eliminates the loss of dissipated power of a resistance control and provides an infinitely smooth regulated power output to the lighting load.

Wound-rotor motors can be started smoothly and operate at speeds commensurate with the load when a saturable-core reactor, connected in series with the motor, provides the control. This completely eliminates the maintenance of costly grid resistors and drum controllers.

Although details of design will vary with the manufacturer, usually three basic styles are available: small, medium, and large. The smaller sizes are constructed with two ac coils on a common core, with the dc control winding on the center leg of the core. Medium-sized reactors normally have two coils, while larger-size reactors normally utilize four coils.

TRANSFORMER GROUNDING

Grounding is necessary to remove static electricity in transformers and also as a precautionary measure in case the transformer windings accidentally come in contact with the core or enclosure. All should be grounded and bonded to meet NE Code requirements and also local codes, where applicable.

The tank of every power transformer should be grounded to eliminate the possibility of obtaining static shocks from it or being injured by accidental grounding of the winding to the case. A grounding lug is provided on the base of most transformers for the purpose of grounding the case and fittings.

The NE Code specifically states the requirements of grounding and should be followed in every respect. Furthermore, certain advisory rules recommended by manufacturers provide additional protection beyond that of the NE Code. In general, the code requires that separately derived alternating-current systems be grounded as stated in Article 250-26.

For additional information on transformers and their application consult the book, *Handbook of Power Generation: Transformers and Generators*, John E. Traister, published by Prentice-Hall, Inc. This book touches briefly on theory and then progresses immediately into practical applications that can be used for virtually every possible situation.

14

Electric Motors

The subject of electric motors is so vast that it deserves a complete book for adequate coverage. However, any book dealing with industrial electrical wiring should have at least a basic description of electric motors and their use in industrial applications.

In basic terms, electric motors convert electric energy into the productive power of rotary mechanical force. This capability finds application in unlimited ways from explosionproof, water-cooled motors for underground mining to induced-draft fan motors for power generation; from adjustable-frequency drives for waste and water treatment pumping to eddy-current clutches for automobile production; from direct-current drive systems for paper production to photographic film manufacturing; from rolled-shell shaftless motors for machine tools to large outdoor motors for crude oil pipelines; from mechanical variable-speed drives for woodworking machines to complex adjustable-speed drive systems for textiles. All these and more represent the scope of electric motor participation in powering and controlling the machines and processes of industries throughout the world.

Before covering electric motor operating principles and their applications, certain motor terms must be understood. Some of the more basic ones are as follows:

Style number: Identifies that particular motor from any other. Manufacturers provide style numbers on the motor nameplate and in the written specifications.

Serial data code: The first letter is a manufacturing code used at the factory. The second letter identifies the month, and the last two numbers identify the year of manufacture (e.g., D78 is April 78).

Frame: Specifies the shaft height and motor-mounting dimensions and provides recommendations for standard shaft diameters and usable shaft extension lengths.

Service factor: A service factor (SF) is a multiplier that, when applied to the rated horsepower, indicates a permissible horsepower loading that may be carried out continuously when the voltage and frequency are maintained at the value specified on the nameplate; the motor will still operate at an increased temperature rise.

Nema service factors: Open motors only

Hp	SF	Hp	SF	Hp	SF
1/12	1.40	1/3	1.35	1	1.15
1/8	1.40	1/2	1.25	1-1/2	1.15
1/6	1.35	3/4	1.25	2	1.15
1/4	1.35	—	—	3	1.15

Phase: Indicates whether the motor has been designed for single or three phase. It is determined by the electrical power source.

Degree C ambient: The air temperature immediately surrounding the motor. Forty degrees centigrade is the NEMA maximum ambient temperature.

Insulation class: The insulation system is chosen to ensure the motor will perform at the rated horsepower and service factor load.

Horsepower (hp): Defines the rated output capacity of the motor. It is based on breakdown torque, which is the maximum torque a motor will develop without an abrupt drop in speed.

Revolutions per minute (rpm): The rpm reading on motors is the approximate full-load speed. The speed of the motor is determined by the number of poles in the winding. A two-pole motor runs at an approximate speed of 3450 rpm. A four-pole motor runs at an approximate speed of 1725 rpm. A six-pole motor runs at an approximate speed of 1140 rpm.

Amps: The amperes of current the motor draws at full load. When two values are shown on the nameplate, the motor usually has a dual voltage rating. Volts and amperes are inversely proportional; the higher the voltage the lower the amperes, and vice versa. The higher ampere value corresponds to the lower voltage rating on the nameplate. Two-speed motors will also show two ampere readings.

Hertz (cycles): Just about everything in this country is serviced by 60-Hz alternating current. Therefore, most applications will be for 60-Hz operations.

Volts: Volts is the electrical potential "pressure" for which the motor is designed. Sometimes two voltages are listed on the nameplate, for example 115/230. In this case the motor is intended for use on either a 115- or 230-V circuit. Special instructions are furnished for connecting the motor for each of the different voltages.

kVA code: This code letter is defined by NEMA standards to designate the locked rotor kilovolt-amperes (kVA) per horsepower of a motor. It relates to starting current and selection of fuse or circuit breaker size.

Housing: Designates the type of motor enclosure. The most common types are open and enclosed.

Open drip-proof—ventilating openings so constructed that successful operation is not interfered with when drops of liquid or solid particles strike or enter the enclosure at any angle from 0° to 15° downward from the vertical.

Open guarded—all openings giving direct access to live metal hazardous rotating parts so sized or shielded as to prevent accidental contact as defined by probes illustrated in the NEMA standard.

Totally enclosed—constructed to prevent the free exchange of air between the inside and outside of the motor casing.

Totally enclosed fan-cooled—equipped for external cooling by means of a fan that is an integral part of the motor.

Air-over—must be mounted in the airstream to obtain their nameplate rating without overheating. An air-over motor may be either open or closed.

Explosionproof motors: These are totally enclosed designs built to withstand an explosion of gas or vapor within it, and to prevent ignition of the gas or vapor surrounding the motor by sparks or explosions that may occur within the motor casing.

Hours: Designates the duty cycle of a motor. Most fractional-horsepower motors are marked continuous for around the clock operation at nameplate rating in the rated ambient. Motors marked "one-half" are for half-hour operation ratings, and those marked "one" are for one-hour operation ratings.

The following terms are not found on the nameplate but are important considerations for proper motor selection:

Sleeve bearings: Sleeve bearings are generally recommended for axial thrust loads of 210 lb or less and are designed to operate in any mounting position as long as the belt pull is not against the bearing window. On light-duty applications, sleeve bearings can be expected to perform a minimum of 25,000 hr without relubrication.

Ball bearings: These are recommended where axial thrust exceeds 20 lb. They too can be mounted in any position. Standard- and general-purpose ball-bearing motors are factory lubricated and under normal conditions will require no additional lubrication for many years.

Rigid mounting: A rectangular steel mounting plate that is welded to the motor frame or cast integral with the frame; it is the most common type of mounting.

Resilient mounting: A mounting base that is isolated from motor vibration by means of rubber rings secured to the end bells.

Flange mounting: A special end bell with a machined flange that has two or more holes through which bolts are secured. Flange mountings are commonly used on such applications as jetty pumps and oil burners.

Rotation: For single-phase motors, the standard rotation, unless otherwise noted, is counterclockwise facing the lead or opposite shaft end. All motors can be reconnected at the terminal board for opposite rotation, unless otherwise indicated.

SINGLE-PHASE MOTORS

Split-Phase Motors

Split-phase motors are fractional-horsepower units that use an auxiliary winding on the stator to aid in starting the motor until it reaches its proper rotation speed (see Fig. 14-1). This type of motor finds use in small pumps, oil burners, automatic washers, and other household appliances.

In general, the split-phase motor consists of a housing, a laminated iron-core stator with embedded windings forming the inside of the cylindrical housing, a rotor made up of copper bars set in slots in an iron core and connected to each other by copper rings around both ends of the core, plates that are bolted to the housing and contain the bearings that support the rotor shaft, and a centrifugal switch inside the housing. This type of rotor is often called a squirrel-cage rotor since the configuration of the copper bars resembles an actual cage. These motors have no windings, as such, and a centrifugal switch is provided

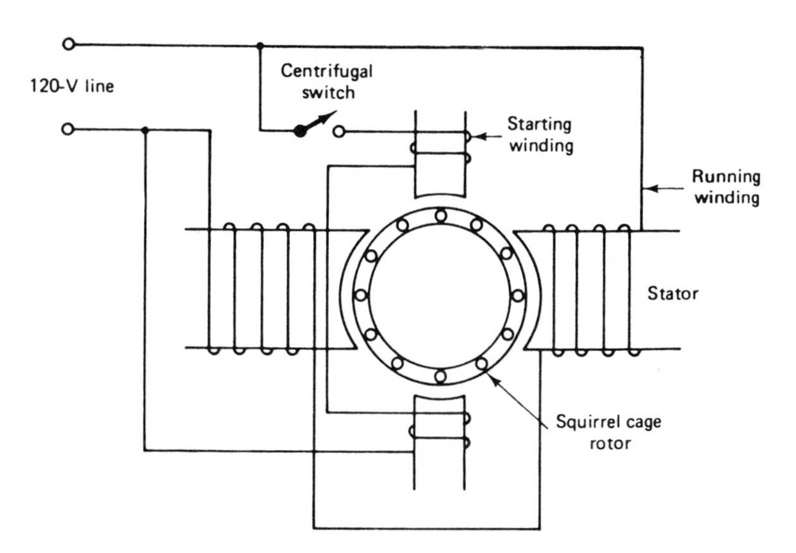

Figure 14-1 Wiring diagram of split-phase motor.

to open the circuit to the starting winding when the motor reaches running speed.

To understand the operation of a split-phase motor, look at the wiring diagram in Fig. 14-1. Current is applied to the stator windings, both the main running winding and the starting winding, which is in parallel with it through the centrifugal switch. The two windings set up a rotating magnetic field, and this field sets up a voltage in the copper bars of the squirrel-cage rotor. Because these bars are shortened at the ends of the rotor, current flows through the rotor bars. The current-carrying rotor bars then react with the magnetic field to produce motor action. When the rotor is turning at the proper speed, the centrifugal switch cuts out the starting winding since it is no longer needed.

Capacitor Motors

Capacitor motors are single-phase ac motors ranging in size from fractional horsepower to about 15 hp. This type of motor is widely used in all types of single-phase applications such as powering machine shop tools (e.g., lathes, drill presses, air compressors, and refrigerators). This type of motor is similar in construction to the split-phase motor, except a capacitor is wired in series with the starting winding, as shown in Fig. 14-2.

The capacitor provides higher starting torque with lower starting current, than does the split-phase motor. Although the capacitor is sometimes mounted inside the motor housing, it is more often mounted on top of the motor, encased in a metal compartment.

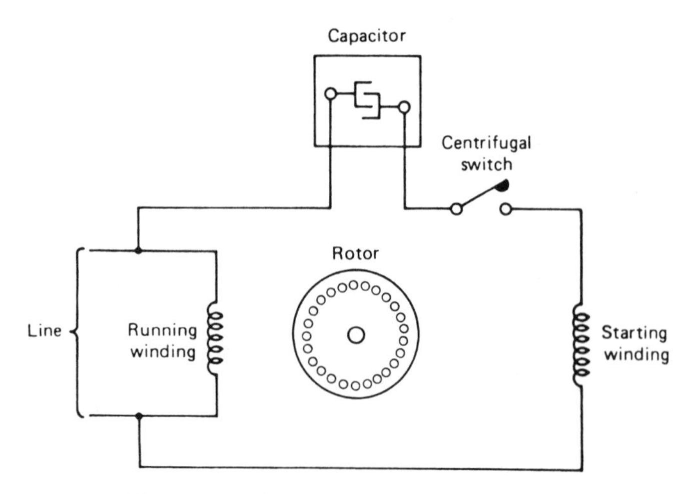

Figure 14-2 Diagram of a capacitor motor.

In general, two types of capacitor motors are in use: the capacitor-start motor and the capacitor start-and-run motor. As the name implies, the former utilizes the capacitor only for starting; it is disconnected from the circuit once the motor reaches running speed, or at about 75% of the motor's full speed. Then the centrifugal switch opens to cut the capacitor out of the circuit.

The capacitor start-and-run motor keeps the capacitor and starting winding in parallel with the running window, providing a quiet and smooth operation at all times.

Capacitor split-phase motors require the least maintenance of all single-phase motors, but they have a very low starting torque, making them unsuitable for many applications. Its high maximum torque, however, makes it especially useful for such tools as floor sanders or in grinders where momentary overloads due to excessive cutting pressure are experienced. It is also used quite frequently for slow-speed, direct-connected fans.

Repulsion-Type Motors

Repulsion-type motors are divided into three groups, including (1) repulsion-start, induction-run motors, (2) repulsion motors, and (3) repulsion-induction motors. The repulsion-start, induction-run motor is of the single-phase type, ranging in size from about 1/10 hp to as high as 20 hp. It has high starting torque and a constant-speed characteristic, which makes it suitable for such applications as commercial refrigerators, compressors, pumps, and similar applications.

The repulsion motor is distinguished from the repulsion-start, induction-run motor by the fact that it is made exclusively as a brush-riding type and does not have any centrifugal mechanism. Therefore, this motor both starts and runs on the repulsion principle. This type of motor has high starting torque and a variable-speed characteristic. It is reversed by shifting the brush holder to either side of the neutral position. Its speed can be decreased by moving the brush holder farther away from the neutral position.

The repulsion-induction motor combines the high starting torque of the repulsion-type and the good speed regulation of the induction motor. The stator of this motor is provided with a regular single-phase winding, while the rotor winding is similar to that used on a dc motor. When starting, the changing single-phase stator flux cuts across the rotor windings and induces currents in them; thus, when flowing through the commutator, a continuous repulsive action on the stator poles is present.

This motor starts as a straight repulsion-type and accelerates to about 75% of normal full speed when a centrifugally operated device connects all the commutator bars together and converts the winding to an equivalent squirrel-cage type. The same mechanism usually raises the brushes to reduce noise and wear. Note that when the machine is operating as a repulsion-type motor, the rotor and stator poles reverse at the same instant, and that the current in the commutator and brushes is ac.

This type of motor will develop four to five times normal full-load torque and will draw about three times normal full-load current when starting with full-line voltage applied. The speed variation from no load to full load will not exceed 5% of normal full-load speed.

The repulsion-induction motor is used to power such items as air compressors, refrigerators, pumps, meat grinders, small lathes, small conveyors, and stokers. In general, this type of motor is suitable for any load that requires a high starting torque and constant-speed operation. Most motors of this type are less than 5 hp.

Universal Motors

This type of motor is a special adaptation of the series-connected dc motor, and it gets the name universal from the fact that it can be connected on either ac or dc and still operate the same. All are single-phase motors for use on 120 or 240 V.

In general, the universal motor contains field windings on the stator within the frame, an armature with the ends of its windings brought out to a commutator at one end, and carbon brushes that are held in place by the motor's end plate, allowing them to have a proper contact with the commutator.

When current is applied to a universal motor, either ac or dc, the current flows through the field coils and the armature windings in series. The magnetic field set up by the field coils in the stator react with the current-carrying wires on the armature to produce rotation.

Universal motors are used on such household appliances as sewing machines, vacuum cleaners, and electric fans.

Shaded-Pole Motor

A shaded-pole motor is a single-phase induction motor provided with an uninsulated and permanently short-circuited auxiliary winding displaced in magnetic position from the main winding. The auxiliary winding is known as the shading coil and usually surrounds one-third to one-half of the pole (see Fig. 14-3). The main winding surrounds the entire pole and may consist of one or more coils per pole.

Applications for this motor include small fans, timing devices, relays, radio dials, or any constant-speed load not requiring high starting torque.

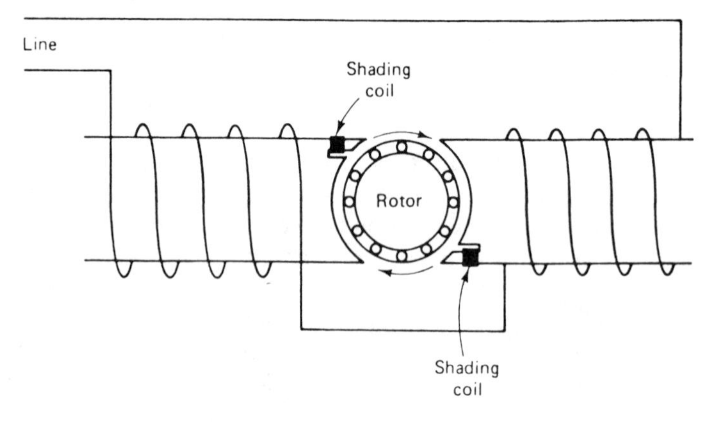

Figure 14-3 Diagram of shaded-pole motor.

POLYPHASE MOTORS

Three-phase motors offer extremely efficient and economical application and are usually the preferred type for commercial and industrial applications when three-phase service is available. In fact, the great bulk of motors sold are standard ac three-phase motors. These motors are available in ratings from fractional horsepower up to thousands of horsepower in practically every standard voltage and frequency. In fact, there are few applications for which the three-phase motor cannot be put to use.

Figure 14-4 Typical three-phase motor.

Three-phase motors are noted for their relatively constant speed characteristic and are available in designs giving a variety of torque characteristics; that is, some have a high starting torque and others, a low starting torque. Some are designed to draw a normal starting current, others, a high starting current.

A typical three-phase motor is shown in Fig. 14-4. Note that the three main parts are the stator, rotor, and end plates. It is very similar in construction to conventional split-phase motors except that the three-phase motor has no centrifugal switch.

The stator shown in Fig. 14-5 consists of a steel frame and a laminated iron core and winding formed of individual coils placed in slots. The rotor may be a squirrel cage or wound-rotor type. Both contain a laminated core pressed onto a shaft. The squirrel-cage rotor is shown in Fig. 14-6 and is similar to a split-phase motor. The wound rotor is shown in Fig. 14-7 and has a winding on the core that is connected to three slip rings mounted on the shaft.

The end plates or brackets are bolted to each side of the stator frame and contain the bearings in which the shaft revolves. Either ball bearings or sleeve bearings are used.

Induction motors, both single-phase and polyphase, get their name from the fact that they utilize the principle of electromagnetic induction. An induction motor has a stationary part, or stator, with windings connected to the ac supply, and a rotation part, or rotor, which contains coils or bars. There is no electrical connection between the stator and rotor. The magnetic field produced in the stator windings induces a voltage in the rotor coils or bars.

Since the stator windings act in the same way as the primary winding of a transformer, the stator of an induction motor is sometimes called the primary. Similarly, the rotor is called the secondary because it carries the induced voltage in the same way as the secondary of a transformer.

Figure 14-5 Stator of a three-phase motor.

The magnetic field necessary for induction to take place is produced by the stator windings. Therefore, the induction-motor stator is often called the field and its windings are called field windings.

The terms primary and secondary relate to the electrical charac-

Figure 14-6 Squirrel-cage rotor. (Courtesy of Westinghouse Electric Corp.)

Figure 14-7 Wound rotor.

teristics and the terms stator and rotor to the mechanical features of induction motors.

The rotor transfers the rotating motion to its shaft, and the revolving shaft drives a mechanical load or a machine, such as a pump, spindle, or clock.

Commutator segments, which are essential parts of dc motors, are not needed on induction motors. This greatly simplifies the design and the maintenance of induction motors as compared to dc motors.

The turning of the rotor in an induction motor is due to induction. The rotor, or secondary, is not connected to any source of voltage. If the magnetic field of the stator, or primary, revolves, it will induce a voltage in the rotor, or secondary. The magnetic field produced by the induced voltage acts in such a way that it makes the secondary follow the movement of the primary field.

The stator, or primary, of the induction motor does not move physically. The movement of the primary magnetic field must thus be achieved electrically. A rotating magnetic field is made possible by a combination of two or more ac voltages that are out of phase with each other and applied to the stator coils. Direct current will not produce a rotating magnetic field. In three-phase induction motors, the rotating magnetic field is obtained by applying a three-phase system to the stator windings.

The direction of rotation of the rotor in an ac motor is the same as that of its rotating magnetic field. In a three-phase motor the direction can be reversed by interchanging the connections of any two supply leads. This interchange will reverse the sequence of phases in the stator, the direction of the field rotation, and therefore the direction of rotor rotation.

SYNCHRONOUS MOTORS

A synchronous polyphase motor has a stator constructed in the same way as the stator of a conventional induction motor. The iron core has slots into which coils are wound, which are also arranged and connected in the same way as the stator coils of the induction motor. These are in turn grouped to form a three-phase connection, and the three free leads are connected to a three-phase source. Frames are equipped with air ducts, which aid the cooling of the windings, and coil guards, which protect the winding from damage.

The rotor of a synchronous motor carries poles that project toward the armature; they are called salient poles. The coils are wound on laminated pole bodies and connected to slip rings on the shaft. A squirrel-cage winding for starting the motor is embedded in the pole faces.

The pole coils are energized by direct current, which is usually supplied by a small dc generator called the exciter. This exciter may be mounted directly on the shaft to generate dc voltage, which is applied through brushes to slip rings. On low-speed synchronous motors, the exciter is normally belted or of a separate high-speed motor-driven type.

The dimensions and construction of synchronous motors vary greatly, depending on the rating of the motors. However, synchronous motors for industrial power applications are rarely built for less than about 25 hp; in fact, most are 100 hp or more. All are polyphase motors when built in the larger sizes. Vertical and horizontal shafts with various bearing arrangements and various enclosures cause wide variations in the appearance of the synchronous motor.

Synchronous motors are used in electrical systems where improvement in power factor is needed or where low power factor is not desirable. This type of motor is especially suited for heavy loads that operate for long periods of time without stopping, such as for air compressors, pumps, and ship propulsion.

The construction of the synchronous motor is well adapted for high voltages, as it permits good insulation. Synchronous motors are frequently used at 2300 V or more. Its efficient slow-running speed is another advantage.

DIRECT-CURRENT MOTORS

A direct-current motor is a machine for converting dc electrical energy into rotating mechanical energy. The principle underlying the operation of a dc motor is called motor action and is based on the fact that, when a wire-carrying current is placed in a magnetic field, a force is exerted on the wire, moving it through the magnetic field. There are three elements to motor action as it takes place in a dc motor

1. Many coils of wire are wound on a cylindrical rotor or armature on the shaft of the motor.
2. A magnetic field necessary for motor action is created by placing fixed electromagnetic poles around the inside of the cylindrical motor housing. When current is passed through the fixed coils, a magnetic field is set up without the housing. Then, when the armature is placed inside the motor housing, the wires of the armature coils will be situated in the field of magnetic lines of force set up by the electromagnetic poles arranged around the stator. The stationary cylindrical part of the motor is called the stator.
3. The shaft of the armature is free to rotate because it is supported at both ends by bearing brackets. Freedom of rotation is assured by providing clearance between the rotor and the faces of the magnetic poles.

Shunt-Wound Direct-Current Motors

In this type of motor, the strength of the field is not affected appreciably by a change in the load, so a relatively constant speed is obtainable. This type of motor may be used for the operation of machines that require an approximately constant speed and impose low starting torque and light overload on the motor.

Series-Wound Direct-Current Motors

In motors of this type any increase in load results in more current passing through the armature and the field windings. As the field is strengthened by this increased current, the motor speed decreases. Conversely, as the load is decreased, the field is weakened and the speed increases, so that at very light loads speed may become excessive. For this reason, series-wound motors are usually directly connected or geared to the load to prevent runaway. The increase in armature current with an increasing load produces increased torque, so the series-wound motor is particularly suited to heavy starting duty and where severe overloads may be expected. Its speed may be adjusted by means of a variable resistance placed in series with the motor, but due to variation with load, the speed cannot be held at any constant value. This variation of speed with load becomes greater as the speed is reduced. Use of this motor is normally limited to traction and lifting service.

Compound-Wound Motors

In this type of motor, the speed variation due to the load changes is much less than in the series-wound motor, but greater than in the shunt-wound motor. It also has a greater starting torque than the shunt-wound motor and is able to withstand heavier overloads. However, it has a narrower adjustable-speed range. Standard motors of this type have a cumulative-compound winding, the differential-compound winding being limited to special applications. They are used where the starting load is very heavy or where the load changes suddenly and violently, as with reciprocating pumps, printing presses, and punch presses.

Brushless Direct-Current Motors

The brushless dc motor was developed to eliminate commutator problems in missiles and spacecraft in operation above the earth's atmosphere. Two general types of brushless motors are in use: the inverter-induction motor, and a dc motor with an electronic commutator.

The inverter-induction motor uses an inverter that uses the motor windings as the usual filter. The operation is square wave, and the combined efficiencies of the inverter and induction motor are at least as high as for a dc motor alone. In all cases, the motors must be designed to saturate so that starting current is limited; otherwise, the transistors or silicon-controlled rectifiers in the inverter will be overloaded. The other type utilizes electronic circuits to control movement of the commutator.

MOTOR ENCLOSURES

Electric motors differ in construction and appearance, depending on the type of service for which they are to be used. Open and closed frames are quite common. In the former enclosure, the motor's parts are covered for protection, but the air can freely enter the enclosure. Further designations for this type of enclosure include drip proof, weather-protected, and splash proof.

Totally enclosed motors, such as the one shown in Fig. 14-4, have an airtight enclosure. They may be fan cooled or self-ventilated. An enclosed motor equipped with a fan has the fan as an integral part of the machine, but external to the enclosed parts. In the self-ventilated enclosure, no external means of cooling is provided.

The type of enclosure to use will depend on the ambient and surround conditions. In a drip-proof machine, for example, all ventilating openings are so constructed that drops of liquid or solid particles falling on the machine at an angle of not greater than 15° from the vertical cannot enter the machine, even directly or by striking and running along a horizontal or inclined surface of the machine. The application of this machine would lend itself to areas where liquids are processed.

An open motor having all air openings that give direct access to live or rotating parts, other than the shaft, limited in size by the design of the parts or by screen to prevent accidental contact with such parts is classified as a drip-proof, fully guarded machine. In such enclosures, openings will not permit the passage of a cylindrical rod 1/2 in. in diameter, except where the distance from the guard to the live rotating parts is more than 4 in., in which case the openings will not permit the passage of a cylindrical rod 3/4 in. in diameter.

There are other types of drip-proof machines for special applications such as externally ventilated and pipe ventilated, which as the names imply are either ventilated by a separate motor-driven blower or cooled by ventilating air from inlet ducts or pipes.

An enclosed motor whose enclosure is designed and constructed to withstand an explosion of a specified gas or vapor that may occur within the motor and to prevent the ignition of this gas or vapor surrounding the machine is designated explosionproof (XP).

Hazardous atmospheres (requiring XP enclosures) of both a gaseous and dusty nature are classified by the NE Code as follows:

- Class I, Group A: atmospheres containing acetylene
- Class I, Group B: atmospheres containing hydrogen gases or vapors of equivalent hazards such as manufactured gas
- Class I, Group C: atmospheres containing ethyl ether vapor
- Class I, Group D: atmospheres containing gasoline, petroleum, naptha, alcohols, acetone, lacquer-solvent vapors, and natural gas
- Class II, Group E: atmospheres containing metal dust

- Class II, Group F: atmospheres containing carbon-black, coal, or coke dust
- Class II, Group G: atmospheres containing grain dust

The proper motor enclosure must be selected to fit the particular atmospheres. However, explosionproof equipment is not generally available for Class I, Groups A and B, and it is therefore necessary to isolate motors from the hazardous area.

MOTOR TYPES

The type of motor will determine the electrical characteristics of the design. An *A* motor is a three-phase, squirrel-cage motor designed to withstand full-voltage starting with locked rotor current higher than the values for a *B* motor and having a slip at rated load of less than 5%.

A *B* motor is a three-phase, squirrel-cage motor designed to withstand full-voltage starting and developing locked rotor and breakdown torques adequate for general application, and having a slip at rated load of less than 5%.

A *C* motor is a three-phase, squirrel-cage motor designed to withstand full-voltage starting, developing locked rotor torque for special high-torque applications, and having a slip at rated load of less than 5%.

Design *D* is a three-phase, squirrel-cage motor designed to withstand full-voltage starting, developing 275% locked rotor torque, and having a slip at rated load of 5% or more.

SELECTING ELECTRIC MOTORS

Each type of motor has its particular field of usefulness. Because of its simplicity, economy, and durability, the induction motor is more widely used for industrial purposes than any other type of ac motor, especially if a high-speed drive is desired.

If ac power is available, all drives requiring constant speed should use squirrel-cage induction or synchronous motors on account of their ruggedness and lower cost. Drives requiring varying speeds, such as fans, blowers, or pumps, may be driven by wound-rotor induction motors. However, if there are machine tools or other machines requiring adjustable speed or a wide range of speed control, it will probably be desirable to install dc motors on such machines and supply them from the ac system by motor-generator sets or electronic rectifiers.

Practically all constant-speed machines may be driven by ac squirrel-cage motors because they are made with a variety of speed and torque characteristics. When large motors are required or when power supply is limited, the wound-rotor motor is used even for driving constant-speed machines. A wound-rotor motor, with its controller and resistance, can develop full-load torque at starting with not more than full-load torque at starting, depending on the type of motor and the starter used.

For varying-speed service, wound-rotor motors with resistance control are used for fans, blowers, and other apparatus for continuous duty, and for cranes, hoists, and other installations for intermittent duty. The controller and resistors must be properly chosen for the particular application.

Synchronous motors may be used for practically any constant-speed drive requiring about 100 hp or over.

Cost is an important consideration where more than one type of ac motor is applicable. The squirrel-cage motor is the least expensive ac motor of the three types considered and requires very little control equipment. The wound-rotor is more expensive and requires additional secondary control. The synchronous motor is even more expensive and requires a source of dc excitation, as well as special synchronizing control to apply the dc power at the correct instant. When very large machines are involved, as, for example, 1000 hp or over, the cost picture may change considerably and should be checked on an individual basis.

The various types of single-phase ac motors and universal motors are used very little in industrial applications, since polyphase ac or dc power is generally available. When such motors are used, however, they are usually built into the equipment by the machinery builder, as in portable tools, office machinery, and other equipment. These motors are, as a rule, especially designed for the particular machine with which they are used.

15

Motor Controls

Electric motors provide one of the principal means of converting electrical energy into mechanical energy for industrial applications; this mechanical energy may then be used to drive modern machine tools, fans or blower units for central heating systems, and other equipment. To be of any use, however, each and every motor must be controlled—if only to start and stop it.

Motor controllers cover a wide range of types and sizes from a simple toggle switch to a complex system consisting of such components as relays, timers, switches, and push buttons. The most common function, however, is the same in every case; that is, to control some operation of an electric motor. These operations may include, but are not limited to, the following:

- Starting and stopping
- Overload protection
- Overcurrent protection
- Reversing
- Changing speed
- Jogging
- Plugging
- Sequence control
- Pilot light indication

A motor-control system can also provide the control for auxiliary equipment such as brakes, clutches, solenoids, heaters and signals, and may be used to control a single motor or a group of motors.

The term *motor starter* is often used in the electrical industry and means practically the same thing as *controller*. Strictly speaking, a motor starter is the simplest form of controller and is capable of starting and stopping the motor and providing it with overload protection.

MANUAL STARTERS

A manual starter is a motor controller whose contact mechanism is operated by a mechanical linkage from a toggle handle or push button, which in turn is operated by hand. A thermal unit and direct acting overload mechanism provides motor-running overload protection. Basically, a manual starter is merely an On/Off switch with overload relays.

Manual starters are used mostly on small machine tools, fans and blowers, pumps, compressors, and conveyors. They have the lowest cost of all motor starters, have a simple mechanism, and provide quiet operation with no ac magnet hum. The contacts, however, remain closed and the lever stays in the On position in the event of a power failure, which causes the motor to automatically restart when the power returns. Therefore, low-voltage protection and low-voltage release are not possible with these manually operated starters. However, this action is an advantage when the starter is applied to motors that should run continuously.

Fractional-horsepower manual starters are designed to control and provide overload protection for motors of 1 hp or less on 120- or 240-V single-phase circuits.

ALTERNATING-CURRENT MAGNETIC STARTERS

Magnetic starters utilize contacts operated by an electromagnetic coil. The contacts close when the coil is energized and open when the coil is de-energized. The coil circuit is energized by some switching device that makes and breaks the circuit to the coil.

Magnetic starters of the full-voltage type apply the full-rated voltage to the motor when the contacts are closed. In addition to the contacts and the electromagnetic operating coil, such a starter includes running overload protective devices that will disconnect the motor from its source of supply when overloaded, up to and including a locked-rotor condition. Across-the-line magnetic starters are made for both single-phase and polyphase motors.

The operating coil of the magnetic starter is a single-phase, two-wire load device commonly connected across two of the supply wires to the motor. A switching device connected in series with the coil provides On/Off control of current flow through the coil. This device is commonly a set of push buttons (Start and Stop) mounted in the front of the starter enclosure or at a location away from the starter to provide remote control of the motor.

Magnetic motor starters are available as two-, three-, or four-pole contactors with one, two, or three overload relays for both single- and three-phase applications, with full-voltage, nonreversing, ac single-speed motors up to 200 hp, 600 V maximum. They are also available for use with small nonreversing dc motors up to and including 1-1/2 hp at 120-V or 2 hp at 240-V.

In selecting this type of starter, consideration should be given to the horsepower rating of the motor, whether single- or three-phase, the coil voltage, the control power transformer, type of enclosure, and the overload relay heaters.

Magnetic reversing across-the-line controllers are frequently used for starting of single- or three-phase motors up to 200 hp, 600 V maximum where the application requires a reversing or plugging action.

When the voltage on the coil of an ac conductor or relay passes through zero its magnetic pull or holding power is zero and the device starts to open. The voltage however is soon effective in the opposite direction and the device is again pulled closed. This operation causes a humming noise in any ac-operated device and a decided chattering noise in a defective unit. The otherwise objectionable chatter is eliminated and the device is kept closed by the use of a shading coil, usually imbedded in the laminated magnetic circuit of the device. The shading coil produces enough out-of-phase flux to provide holding power to maintain the device closed during the short period when the power to main flux is zero. Even with shading coils in use, the air-gap surfaces must be free from dirt and well fitted to avoid objectionable noise. Broken shading coils are ineffective and of course cause noisy operation.

For quiet operation of ac contactors it is necessary to provide well fitted surfaces at the air-gap. Any dirt in this area introduces a greater air-gap when the unit is closed, increases the duty imposed upon the shading coil, and results in noisier operation. To prevent rusting of the fitted surfaces at the air-gap, these devices are often shipped with a small amount of grease or oil on them. This lubrication may cause a "seal" that makes them sticky and sluggish in opening when first put into service. These surfaces should therefore be wiped clean before the units are placed in service.

THERMAL OVERLOAD RELAYS

Thermal overload relays have the inverse time-limit feature, which means the greater the overload, the shorter the time of tripping. They provide excellent protection against overloads and momentary surges but do not protect against short-circuit currents. For protection against the latter, fuses not exceeding four times the motor full-load current, or time-limit circuit breakers set at not more than four times the motor full-load current, or instantaneous trip circuit breakers should be installed ahead of the linestarter. Where fuses are used it is good to use a disconnecting switch as well.

Heaters for thermal relays are made with different current ratings, so that within its limits any starter can be used with different size motors and still afford proper protection by selecting the size of heater that corresponds to the full-load current of the motor being used. The ones in Fig. 15-1 are typical of those supplied by manufacturers. The solder pot, which is the heat-sensitive element, is an integral part of the thermal unit. It provides accurate response to overload current yet prevents nuisance tripping. The heat-winding is the heat-producing element. It is permanently joined to the solder pot so proper heat transfer is always ensured and there is no chance of misalignment in the field. In general, the ampere rating of the heater should be approximately 120% of the motor full-load current.

A further adjustment is possible by means of a calibration lever on some types of relays. When set at 100%, the current stamped on the heater will just trip the starter after several minutes. For tripping at a smaller current, move the level toward 90; or for a larger current, move the lever toward 120.

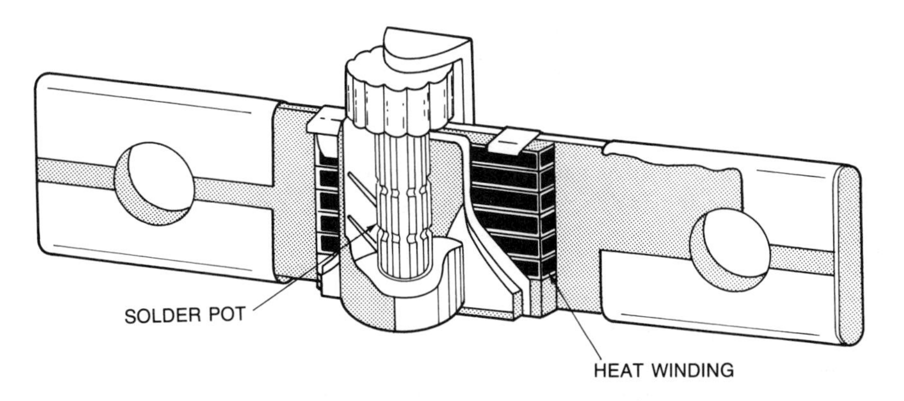

Figure 15-1 Example of heaters for thermal relays.

MAGNETIC REVERSING STARTERS

Across-the-line magnetic reversing motor starters are used for full-voltage frequent starting of single-phase or polyphase motors up to about 200 hp, 600-V maximum, where the application requires reversing or plugging operation.

Three-phase, squirrel-cage motors are particularly suited to reversal of rotation since reversing the connection of any two of the motor feeds will cause a change in rotation. This reversal of line connection is normally done by using two separate contactor assemblies—one for the forward rotation and the other for reverse rotation. Both contactor assemblies are mounted in a single enclosure or cabinet. Reversing starters are used for starting, stopping, and reversing three-phase, squirrel-cage motors; for primary reversing of wound-rotor motors; and for some single-phase applications. These starters are available with or without running overload protection, and they are also available in combination-starter form; that is, with either a fusible or nonfusible On/Off disconnect switch.

Open reversing controllers can also be furnished with either horizontal or vertical mechanical interlocking. Momentary-contact push button control devices provide complete under-voltage protection for the load circuit. Thermal overload relays, when used, provide complete overcurrent protection.

Manual reversing starters (employing two manual starters) are also available. As in the magnetic version, the forward and reverse switching mechanisms are mechanically interlocked, but since coils are not used in the manually operated equipment, electrical interlocks are not furnished.

ALTERNATING-CURRENT REDUCED-VOLTAGE MOTOR STARTERS

Any three-phase motor can be connected directly to the full voltage for which it is rated without doing damage to the motor. However, the in-rush of starting current on full voltage, which can be as much as eight (or more) times the normal running current, can cause severe voltage disturbances in the electrical system from which the motor is supplied. In addition, the shock of the starting torque of very large motors might damage the driven load in some cases. As a result, it may be necessary to start a motor at reduced voltage to prevent or minimize any objectionable effects on the load or the supply system. Starters used for such cases are called reduced voltage starters.

Reduced voltage starters, both manual and automatic, have a broad range of application. Reduced voltage starters of the autotransformer type find extensive application where the size or design of the motor, or restrictions of the supply circuit, require starting on reduced voltage. The autotransformer-type starter provides greater starting torque per ampere starting current drawn from the line than any other type of reduced-voltage motor starter.

One way to start a motor at reduced voltage is to connect a resistor in each conductor to the motor. The resistor limits the current and reduces the voltage applied to the motor windings. There are both manual- and automatic-type resistance starters that first connect resistance in the motor circuit to reduce the voltage to the motor and hold down starting current. Once the motor comes up to a certain speed, the resistance is cut out of the circuit and full voltage is applied to the motor. In manual units, a handle is moved from the Start to the Run position to bring the motor up to speed and normal operation. Magnetic contactors cut out the resistance in the automatic (magnetic) resistance starters.

Probably the most widely used type of reduced-voltage starter at this time is the autotransformer or compensator type. This starter limits the starting current with higher starting torque than other reduced-voltage starters without the energy loss of resistor starters. As with resistance starters, autotransformer starters are available in both manual and automatic types. In the manual type, a handle is first moved to the Start position, held a few seconds, and then moved to the Run position. In the Start position, tap connections are made on an autotransformer assembly in the unit to apply to the motor a voltage less than the full circuit voltage. When the motor has come up to a certain speed, the autotransformer winding is cut out and full voltage is applied to the motor. In the automatic type of autotransformer starter, the switching operations from reduced voltage to full voltage are made by magnetic contactors operated by timing devices.

Autotransformer-type starters are normally rated from 5 to 250 hp. Two taps are provided for 65% and 80% of line voltage in most cases. However, for use on motors above 50 hp, taps are provided for 50, 65, and 80% of the line voltage, giving respective line currents equal to 25, 42, and 64% of the full-voltage starting current.

When selecting an autotransformer starter, the following information should be researched:

1. Motor voltage
2. Start-up torque required
   Caution: The transition from start to run is open (momentary loss of current to motor) with the manual autotransformer starter. If

closed transition is required (power is continuously applied from start to run) use a magnetic form of starter.

3. Heater selection
4. Enclosure type: NEMA 1 or NEMA 3R
5. Shallow mount or with conduit access box
6. Modifications, including combination forms

Common uses for this type of starter include starting duty for blowers, conveyors, and pump motors in connection with automatic pilot devices such as limit switches and pressure switches.

When selecting magnetic reduced-voltage starters, the following should be given consideration:

1. Operation duty cycle
2. Voltage and type of motor to be controlled
3. Overload relay heater selection
4. Type of enclosure
5. Modifications
6. Combination form required

PART-WINDING STARTERS

Part-winding magnetic starters (Fig. 15-2) are sometimes referred to as increment starters. These starters are commonly used to control motors driving light or low-inertia loads, such as air conditioning compressors, refrigerators, compressors, pumps, fans, and blowers. This method of starting has limitations on the type of load that can be accelerated on the first point. Inrush current is limited to an average of approximately 65% of across-the-line starting current, depending upon the use of either two thirds or one half of the motor winding on start.

Part-winding magnetic starters are available in NEMA sizes 1PW to 5PW, Types 1, 4, and 12, as well as open forms.

When selecting this type of motor, the following should be given consideration:

1. Type of motor to be controlled (must be reconnectable, separate, isolable windings suitable for use with part-winding starter)
2. Start-up torque requirements
3. Heater selection
4. Enclosure
5. Modifications

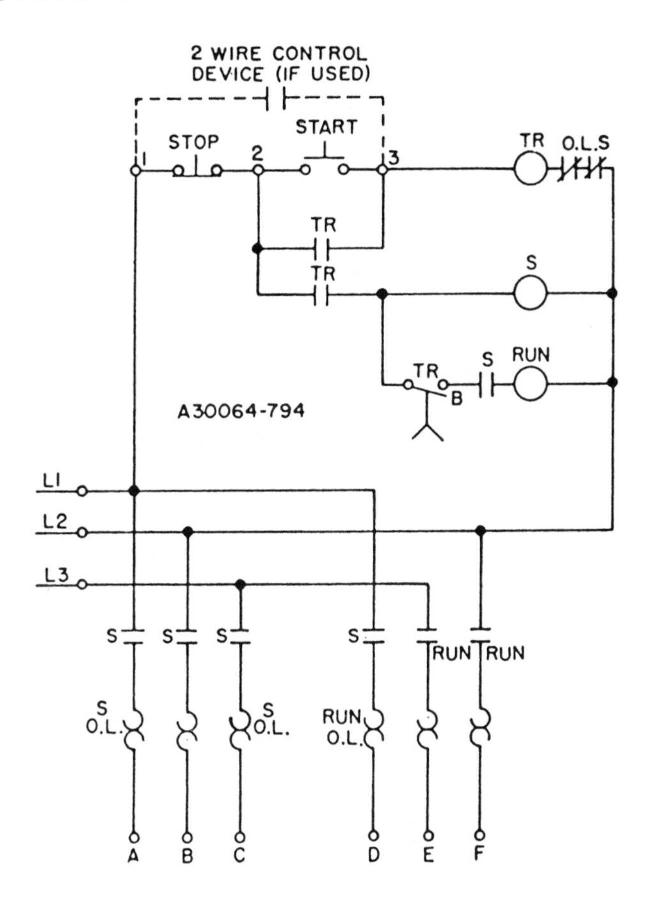

Elementary diagram of sizes 1, 2, 3 & 4,
2 step part winding motor starter

Figure 15-2 A part-winding magnetic starter. (Courtesy of Square D Co.)

WYE-DELTA MAGNETIC STARTERS

Wye-delta magnetic starters are for use with low starting torque applications such as fans, compressors, and conveyors driven by motors capable of being connected in wye and in delta. Wye-delta starting provides a low inrush current, which results in low starting torque. When the motor windings are connected in wye during starting, each winding has 58% full voltage. Automatically reconnecting to delta on run applies full voltage to each winding.

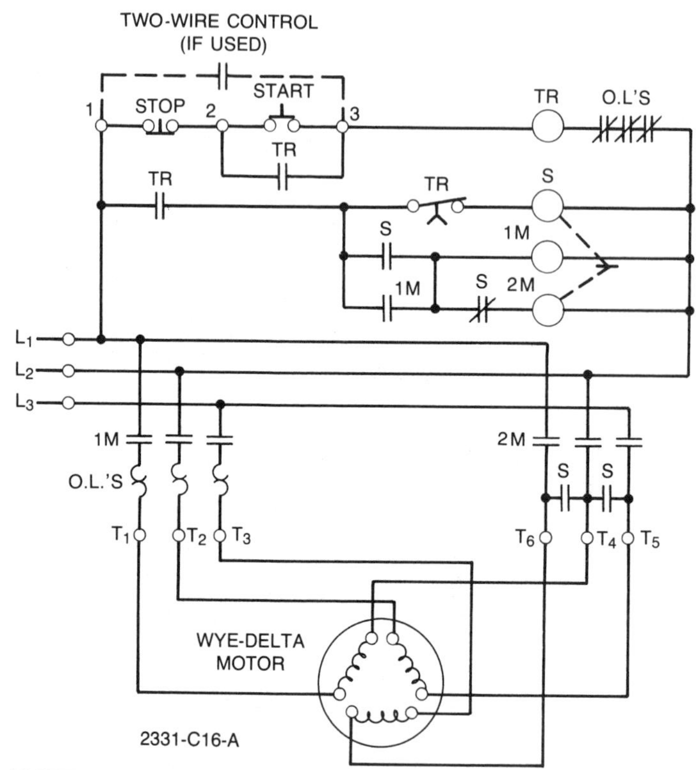

Figure 15-3 Wye-delta magnetic-starter circuit. (Courtesy of Square D Co.)

Either open-transition forms or closed-transition forms are available in open baseplate forms in general-purpose, and watertight or industrial-use enclosures. See Fig. 15-3.

When selecting this type of starter, determine the following:

1. Type of motor to be controlled (must be nameplate rated for starting with wye-delta starters)
2. Start-up torque required
3. Voltage and horsepower of motor
4. Combination form required
5. Overload relay heater selection
6. Modifications

MAGNETIC PRIMARY-RESISTOR STARTERS

Resistance-type starters (Fig. 15-4) are sometimes applied on network distribution systems where power companies' regulations require that the circuit not be opened during the transition from reduced voltage to full voltage. They are particularly desirable to avoid sudden mechanical shocks to the driven load.

Automatic reduced-voltage starters are well adapted for geared or belted drives where a sudden application of full-voltage torque must be avoided.

The inrush current is limited to approximately 80% of the full-voltage locked-rotor value and approximately 64% of the full-voltage locked-rotor starting torque is produced.

General start duty NEMA Class 116 resistors or high-inertia start duty NEMA Class 156 resistors are both available in this model of motor starter.

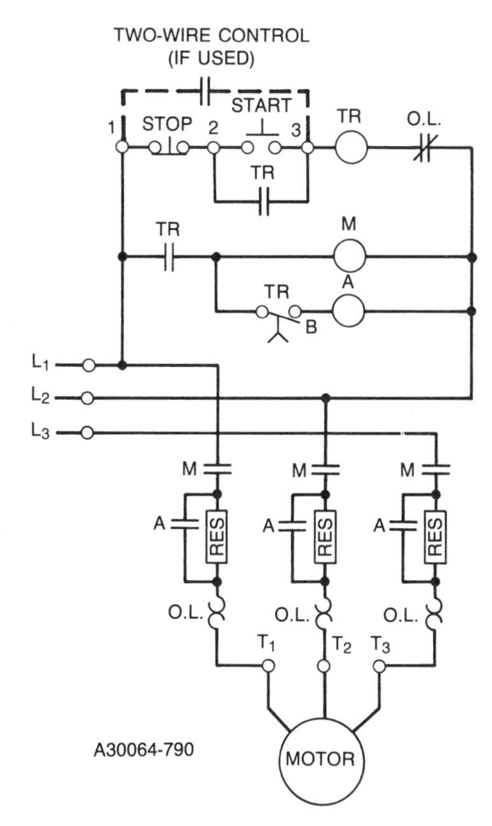

Figure 15-4 Resistance-type starter. (Courtesy of Square D Co.)

The following should be considered when selecting starters:

1. Start duty (torque requirement)
2. Operational voltage and frequency
3. Horsepower of motor
4. Combination form required
5. Overload relay heater selection
6. Enclosure selection

SYNCHRONOUS MOTOR CONTROLS

Introduction

Synchronous motors are used in electrical systems where there is need for improvement in power factor or where a low power factor is not desirable. This type of motor is especially adapted to heavy loads that operate for long periods of time without stopping, such as for air compressors, pumps, and ship propulsion.

The construction of the synchronous motor is well adapted for high voltages, as it permits good insulation. Synchronous motors are frequently used on 2300 V or more. Their efficient slow-running speed is another advantage.

Synchronous polyphase motors have a stator constructed in the same way as the stator coils of induction motors. These are in turn grouped to form a three-phase connection, and the three free leads are connected to a three-phase source. Frames are equipped with air ducts, which help cool the windings, and coil guards, which protect the winding from damage.

The rotor of a synchronous motor carries poles, which project toward the armature and are called salient poles. The coils are wound on laminated pole bodies and connected to slip rings on the shaft. A squirrel-cage winding for starting the motor is embedded in the pole faces.

The pole coils are energized by direct current, which is usually supplied by a small dc generator called the exciter. This exciter may be mounted directly on the shaft to generate dc voltage, which is applied through brushes to slip rings. On low-speed synchronous motors, the exciter is normally belted or of a separate high-speed motor-driven type.

The dimensions and construction of synchronous motors vary greatly, depending on the rating of the motors. However, synchronous motors for industrial power applications are rarely built for less than

25 hp or so. In fact, most are 100 hp or more. All are polyphase motors when built in this size. Vertical and horizontal shafts with various bearing arrangements and various enclosures cause wide variations in the appearance of the synchronous motor.

One distinguishing feature of the synchronous motor is that it runs without slip at the synchronous speed determined by the frequency and the number of poles it has.

Starting Synchronous Motors

Controls used on synchronous motors have two basic functions: to start the motor, and then to bring it up to synchronous speed by exciting the dc field.

Starting a synchronous motor may be accomplished by any of the across-the-line starters, autotransformers, resistance-type motor starters, or other method ordinarily used for induction motors provided their capacity is adequate for the motor being started. However, control of the dc field must also be provided to bring the motor up to synchronous speed.

MULTISPEED ALTERNATING-CURRENT MOTOR STARTERS

Both squirrel-cage and wound-rotor inductive, and occasionally synchronous ac motors, may be arranged with windings that provide two or more speeds. Two-speed motors may have two separate windings, or a single winding capable of rearrangement or pole changing. Four-speed motors usually have two 2-speed windings. In any case, the different speeds require different switching set-ups.

Wound-rotor and synchronous motors require switching or pole changing in both primary and secondary (or field) windings. Multispeed motor starters may consist of manually operated drum switches or of magnetic contactors. Standards for connections and markings have changed over the years and have varied with different motor manufacturers. Therefore, when servicing multispeed controllers, actual connections and markings should be obtained from the wiring diagram or from the motor-connection plate.

Line-voltage type multispeed starters are designed to control separate- and reconnectable-winding squirrel-cage motors to operate at two, three, or four different constant speeds, depending upon their construction. The use of an automatic starter and proper control station permits greater operating efficiency and offers protection to both motor and machine against improper sequencing or too rapid speed change. Protection against motor overload in each speed is provided.

Separate winding-type motors have a winding for each speed required. This motor construction is slightly more expensive, but the controller is relatively simple, and a wide variety of speeds can be selected.

Consequent pole-type motors have a single winding for two speeds. Extra winding taps are brought out to permit reconnection for a different number of stator poles. Although the motor is less expensive, the controller is more complicated, and the speed range is limited to a 2:1 ratio; such as 600–1200, and 900–1800.

STARTING AND SPEED-REGULATING RHEOSTATS AND CONTROLLERS

Rheostats of the face-plate type with self-contained resistors are used in conjunction with manual or magnetic primary control for starting and for speed regulation by secondary control of ac wound-rotor motors ranging from 1/4 to 25 hp. For reversing service and heavier duty applications, drum or drum-contactor controllers are used with separately mounted resistors; the standard drum controller being used for ratings from 1/2 to 100 hp and for the heaviest duty applications, the drum contactor type is used for ratings from 2 to 300 hp.

METHODS OF DECELERATION

There are many times during actual use where one wants to control a motor in ways other than normal starting and stopping. Some of the most used operations include plugging, braking, and jogging.

Plugging

Plugging is an operation in which the connection to a motor is quickly reversed for an instant, causing reverse torque on the motor to bring it to an abrupt stop. This technique may be used either for quick stopping or rapid reversing of motors.

A special type of switch known as a "plugging switch" or a "zero-speed switch" is used to provide plugging. A plugging switch is a centrifugally operated device mounted on the motor shaft where it can sense when the motor rotation has ceased and then will open the reverse contactor. Such switches are often installed on machine tools that are required to come to an abrupt stop at some point in their cycle of operation to prevent damage to the machine or the work itself. Plugging is also used in processes where machines must come to a complete stop before the next phase of operation is begun; thus, the reduced stopping time saves production time.

In actual operation, the shaft of the plugging switch is mechanically connected to the motor shaft, or possibly a shaft on the driven machine, but it is usually best to avoid the latter connection if at all possible. When mounted in this manner, the motion of the motor is transmitted to the switch contacts by a centrifugal mechanism or a magnetic induction arrangement within the switch. The contacts are wired to the reversing motor, which controls the motor. Acting as a link between the motor and reversing starter, the switch allows the starter to apply just enough "reverse" power to bring the motor to a quick stop.

Jogging

The jogging circuit shown in Fig. 15-5 is primarily used when machines must be operated momentarily for inching, such as in a machine tool set up for maintenance. The jog circuit energizes the starter only when the Jog button is depressed, thereby giving the machine operator instantaneous control of the motor drive. In the circuit under consideration, when the Jog button is depressed, the control relay is bypassed,

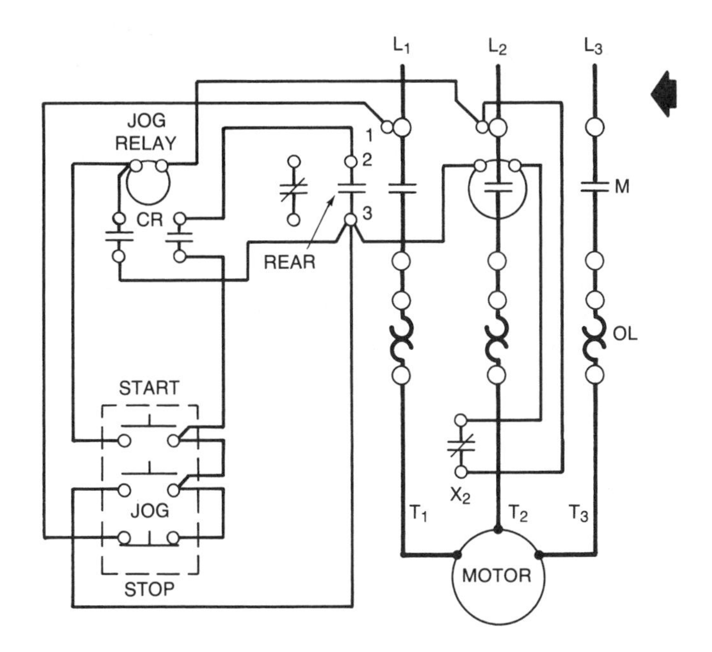

Figure 15-5 A typical jogging circuit.

and the main contactor coil is energized solely through the Jog button; when the Jog button is released, the contactor coil releases immediately. Pushing the Start button closes the control relay, and it is held in by its own normally open contacts. The main contactor coil is in turn closed by another set of normally open contacts on the control relay and is held in the On position.

Electric Brakes

Electric motor brakes are composed of a linkage of moving parts and a friction member. These are applied to hold a load or retard the normal rotation of some machines, or both. Different types are available depending on the power supply, ratings, and the application requirements.

PROGRAMMABLE CONTROLLERS

Programmable controllers may be used in place of conventional controls; they include relays offering faster start up, decreased start-up costs, quick program changes, fast troubleshooting, and up-to-date schematic diagram printouts. They also offer additional benefits such as precision digital timing, counting, data manipulation, remote inputs and outputs, redundant control, supervisory control, process control, management-report generation, machine cycle, controller self-diagnostics, and dynamic graphic displays. Applications include machine tool, sequential, process, conveyor, batching, and energy-management control applications.

16

Industrial Lighting

Lighting layouts for industrial buildings should be designed to provide the highest visual comfort and performance that is consistent with the type of area to be lighted and the budget provided. However, since individual tastes and opinions vary, there can be many solutions to any given lighting problem. Some of these solutions can be commonplace, while others will show imagination and resourcefulness.

The data presented in this chapter are designed to give the reader a basic knowledge of lighting design; that is, how to calculate lighting requirements, lay out lighting schemes, and select the proper lamp sources and lighting fixtures for given areas. The information is by no means complete, but it should serve as a starting point for anyone required to design lighting arrangements for building construction.

LAMP SOURCES

Electric lamps are made in thousands of different types and colors, from a fraction of a watt to over 10 kW each, and for practically any conceivable lighting application.

Incandescent filament lamps, for example, consist of a sealed glass envelope containing a filament that produces light when heated to incandescence (white light) by its resistance to a flow of electric current. This type of light source is relatively inexpensive to install, is not greatly affected by ambient temperatures, is easily controlled as to direction and brightness, and gives a high color quality. Incandescent lamps, as compared to other lamps, are less efficient and result in a higher operating cost per lumen. More heat is produced per lumen than electric-discharge lighting, causing the need for a larger air-conditioning system which in turn increases operating cost.

The quartz-iodine tungsten-filament lamp is similar to the basic incandescent lamp except that the glass envelope contains an iodine vapor, which prevents the evaporation of the tungsten filament. This increases the normal life to about twice that of a normal incandescent lamp.

Fluorescent lighting has a high efficiency as compared to incandescent lighting. To illustrate this fact, the average 40-W inside-frosted incandescent lamp delivers approximately 470 initial lumens, while a cool-white fluorescent of the same wattage delivers over 3200 initial lumens. This power efficiency not only saves on the cost of power consumed, but also lessens the heat produced by lamps, which in turn reduces air-conditioning loads. Further, fluorescent lighting provides a linear source of light, long lamp life, and a means of relatively low surface luminance. However, the initial installation cost is normally higher due to the required auxiliary equipment (ballast, and so on). Also, fluorescent lighting is temperature and humidity sensitive, produces radio interference, and does not lend itself to critical light control.

The most popular lamp types are cool white and deluxe warm white. The cool-white lamp is often selected for offices, factories, and commercial areas where a psychologically cool working atmosphere is desirable. It is also one of the most efficient fluorescent lamps manufactured today.

Deluxe warm-white lamps produce a more flattering color to the complexion; the color is very close to incandescent in that they impart a ruddy or tanned hue to the skin. They are generally recommended for application in homes and for commercial use where flattering effects on people and merchandise are considered important. Whenever a warm social atmosphere is desirable, this is the color of fluorescent lamp to use.

High-Intensity Discharge Lamps

The term *high-intensity discharge lamps* describes a wide variety of lighting sources. Their common characteristic is that they consist of gaseous discharge arc tubes, which, in the versions designed for lighting, operate at pressures and current densities sufficient to generate desired quantities of radiation within their arcs alone.

Mercury vapor lamps contain arc tubes, which are formed of fused quartz. This has resulted in great improvements in lamp life and maintenance of output through life. These arcs radiate ultraviolet energy as well as light, but the glass used in the outer bulbs is generally of a heat-resisting type that absorbs most of the ultraviolet. Some mercury lamps have outer bulbs that are internally coated with fluorescent materials, which, when activated by the ultraviolet, emit visible energy at wavelengths that modify the color of light from the arc. The General Electric deluxe-white mercury lamps, for example, have color characteristics well suited to many commercial lighting applications that could not have been considered for mercury until recently. General-lighting mercury lamps are available in wattages from 50 to 3000.

Multivapor lamps generate light with more than half again the efficiency of the mercury arc, and with better color in the bargain.

In 1965, General Electric introduced the Lucalox lamp, which has the highest light-producing efficiency of any commercial source of white light. This lamp was made possible by the invention of a means of effectively sealing metal ends and electrodes to a tube of Lucalox ceramic in a combination that could withstand temperatures and corrosive effects produced by intensely hot vapors of the alkaline metals. The arc is principally made of metallic sodium, which yields much better color quality and compactness, with substantially higher luminous efficacy than has been available before for white light.

The outer bulbs of high intensity discharge lamps are designed to provide, as nearly as possible, optimum internal environments for arc-tube performance. For example, the rounded shapes labeled *E* and *BT* in the sketches in Fig. 16-1 were devised to maintain uniform temperatures of the bulb walls for better performance of phosphor coatings. The E-bulb improves manufacturing efficiency and eliminates the clear bulb-end on phosphor-lined bulbs.

In some cases, special considerations dictate the bulb shape. The R and PAR contours have been selected to achieve desired directional distribution of light. Some of the smaller T-bulbs are made of highly specialized glasses, which are more economically formed in these simple contours.

Most of the general contours of high intensity discharge lamps are shown in Fig. 16-1, with verbal descriptions of the code used for the shapes. The complete description of a bulb also includes a number that represents the maximum diameter of the bulb in eighths of an inch. The E-37 bulb, therefore, is elliptical in shape and 4-5/8 in. in diameter at its widest point; the R-80 is a reflector bulb with 10-in. maximum diameter.

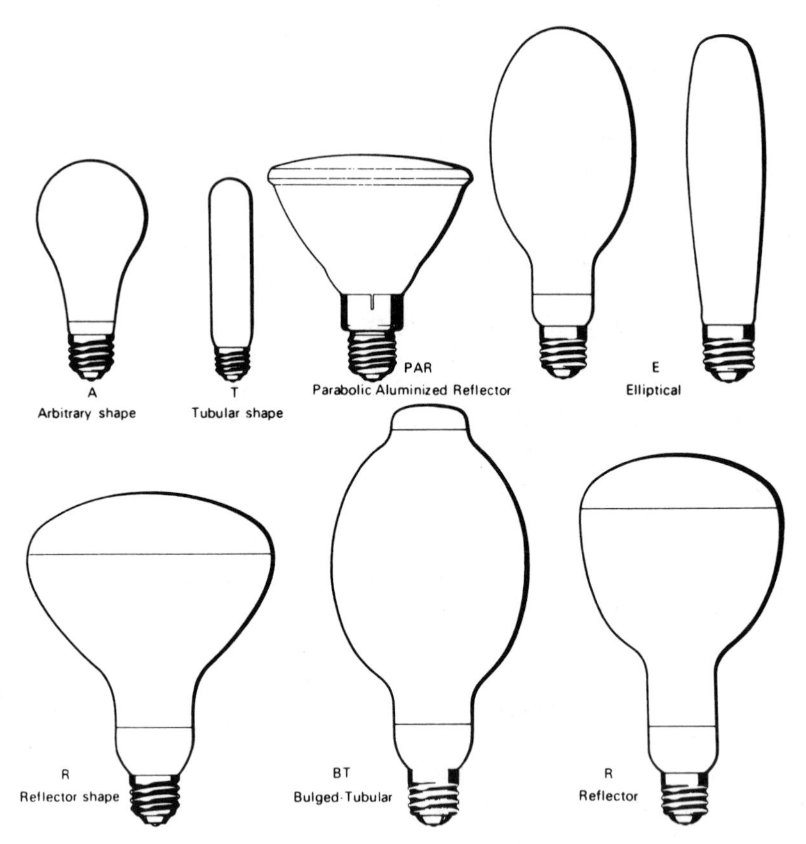

Figure 16-1 The general contour of high-intensity discharge lamps.

Mercury Lamps

A typical mercury lamp consists of the parts schematically illustrated in Fig. 16-2, enclosed in an outer bulb made of borosilicate glass, which can withstand high temperatures, and which is resistant to thermal shocks such as those created when cold raindrops strike a hot bulb. The outer bulb contains a small quantity of nitrogen, an inert gas; this atmosphere maintains internal electrical stability, provides thermal insulation for the arc tube, and protects the metal parts from oxidation. The quartz arc tube contains a small quantity of high-purity mercury, and a starting gas, argon.

Most mercury lamps operate on ac circuits, and the ac-circuit ballast usually consists of a transformer to convert the distribution voltage of the lighting circuit to the required starting voltage for the lamp, and inductive or capacitive reactance components to control lamp current and, in some ballasts, to improve power factor.

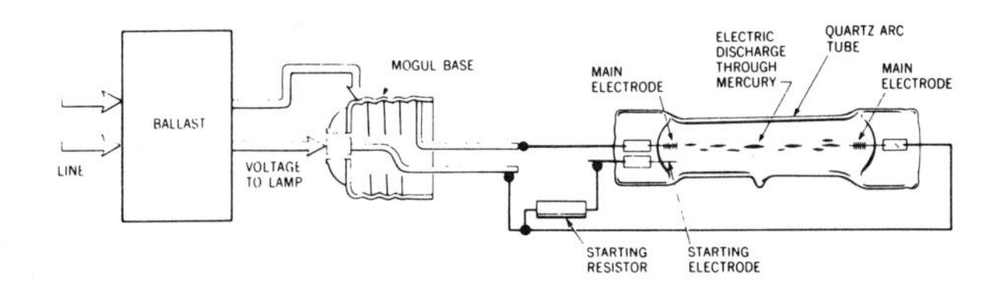

Figure 16-2 Basic parts of a typical mercury lamp.

Most mercury lamps start and operate equally well in any burning position. However, light output and maintenance of output through life generally are slightly higher with vertical than with horizontal operation.

The operating life of mercury lamps is very long, which accounts for much of their popularity in recent years. Most general lighting lamps of 100 to 1000 W have rated lives in excess of 24,000 hr, while the 50-, 75-, and 100-W lamps with medium screw bases are rated at 10,000 hr. (The life rating is the time required to burn out half the lamps in a large sampling.) Ratings are based on operation with properly designed ballasts, with five or more burning hours per start. More frequent starting may reduce life somewhat.

Multivapor Lamps

Multivapor lamps are quite similar in physical appearance to conventional clear mercury lamps. The major differences in internal construction and appearance can be seen by comparing the sketch in Fig. 16-3 with the one in Fig. 16-2. At the present time, two sizes of multi-

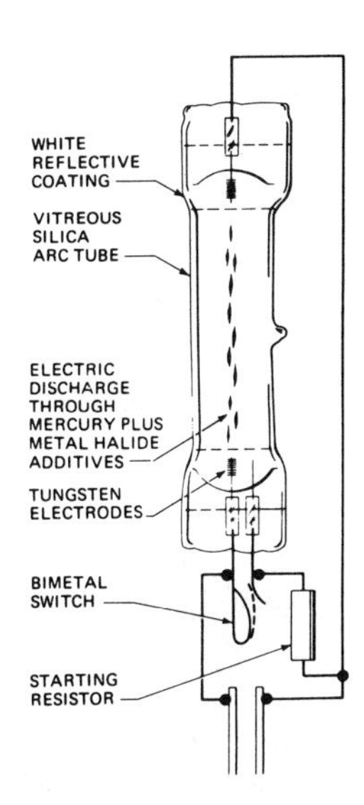

Figure 16-3 Internal components of a multivapor lamp.

vapor lamps are available: 400 and 1000 W. The 1000-W size has been used most widely, largely because it is capable of delivering so much light from a single luminaire. Applications for multivapor to date include industrial lighting, street lighting, building floodlighting, and the floodlighting of several major stadiums to provide the amount and color quality of light needed for color telecasts, without resorting to huge increases in power requirements.

Lucalox Lamps

The construction, operation, and radiation characteristics of Lucalox lamps are quite unlike those of the other high intensity discharge lamps. As the sketch in Fig. 16-4 illustrates, the essential internal components are fewer. The sketch cannot, however, illustrate the tremendous steps in material and processing technology that made this simplification possible. The result has been the most efficient source of white light ever made. The two wattages of Lucalox lamps available at this writing are 275 and 400; they produce light with an efficacy of 100 and 115 lumens per watt (lm/W), respectively. This compares with values of about 80 lm/W for fluorescent, 50 lm/W for mercury, and 15–20 lm/W for incandescent in the types used in commercial and industrial lighting.

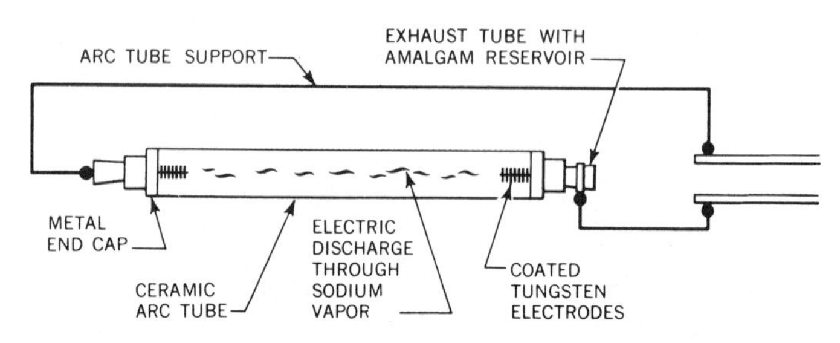

Figure 16-4 Basic components of a Lucalox lamp.

Discharge Lamps for Special Applications

In addition to the more common mercury, multivapor, and Lucalox lamps, which are generally used for lighting purposes, there are a number of types of discharge lamps for special applications. Some of these include the following.

Sunlamps and black light lamps. Many conventional mercury lamps are good sources of "black light" or near ultraviolet.

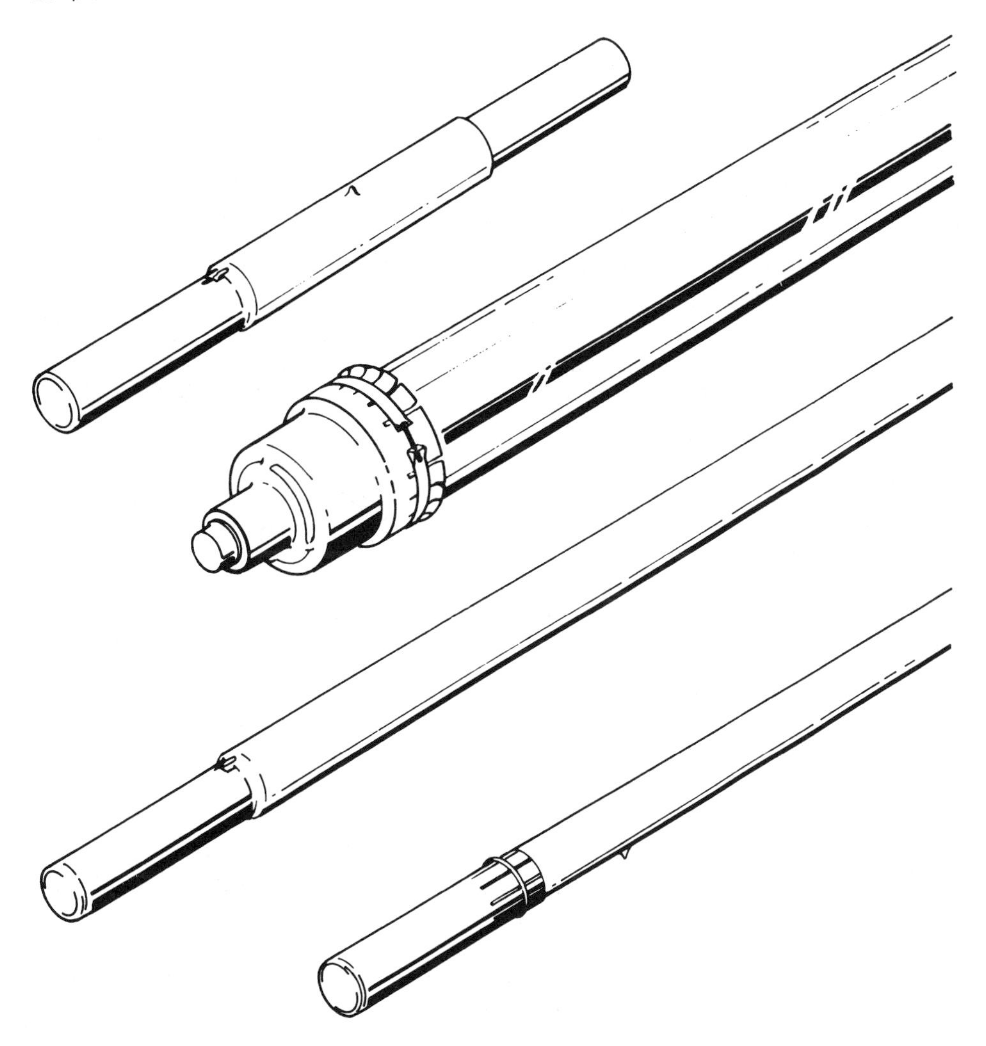

Figure 16-5 Typical uviarc lamps.

Uviarc lamps. Uviarc lamps (Fig. 16-5) are tubular mercury lamps designed and manufactured for use in diazo printing machines for processing engineering drawings. They are also used for copyboard lighting in some printing operations, for simulated weather tests, and for some special chemical processes. Wattages range from 250 to 7500. Some of the lamps have jacketed arc tubes for special performance or radiation characteristics.

Tubular reprographic lamps. A series of tubular mercury lamps (Fig. 16-6) with quartz arc tubes is made in wattages from 250 to 1440; bulbs range from T-3 to T-7 in diameter, 4-1/2 to 26 in. in length.

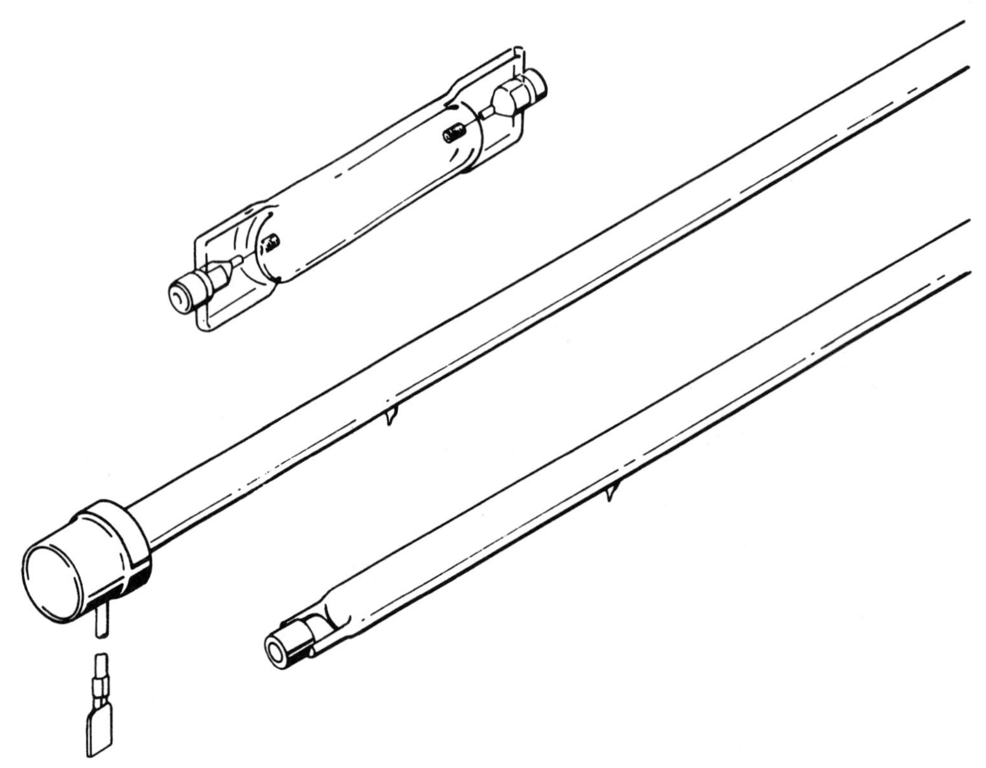

Figure 16-6 Tubular reprographic lamps.

These lamps are primarily designed as ultraviolet radiation sources for common office copying and duplicating machines. They are also used as light sources in making lithographic plates.

Capillary mercury lamps. The AH-6, BH-6, and FH-6 lamps (Fig. 16-7) are highly specialized 1000-W mercury lamps with quartz arc tubes only 1/4 in. × 3-1/4 in. in size, including their brass base sleeves. Actual arc dimensions are about 1.5 × 20–25 mm, producing extremely high luminances (300–360 candelas per square millimeter). They are very potent and concentrated sources of both visible and ultraviolet radiation. The AH-6 operates with forced water cooling, while the BH-6 and FH-6 are air cooled. The lamps are used for a number of specialized applications, including the following: manufacture of color television

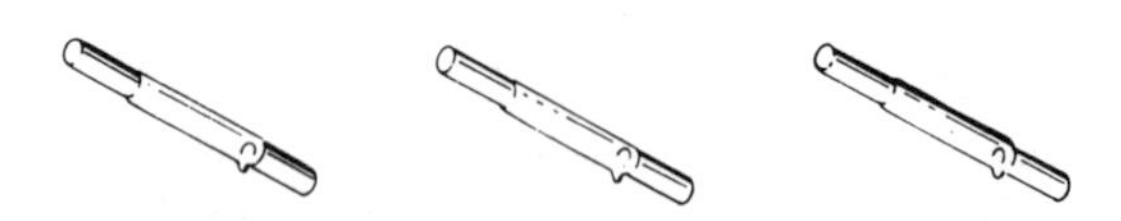

Figure 16-7 Typical capillary mercury lamps.

picture tubes, measuring "ceiling height" at airports, catalyzing chemical reactions by radiation, instrumentation in wind tunnels.

Xenon compact arcs. Three xenon arc lamps (Fig. 16-8) are currently available. Two are 500-W lamps, featuring a compact arc only 2.5 mm long, and one 5000-W lamp has a 7.2-mm arc. They are designed for operation on dc circuits in enclosed equipment. One of the

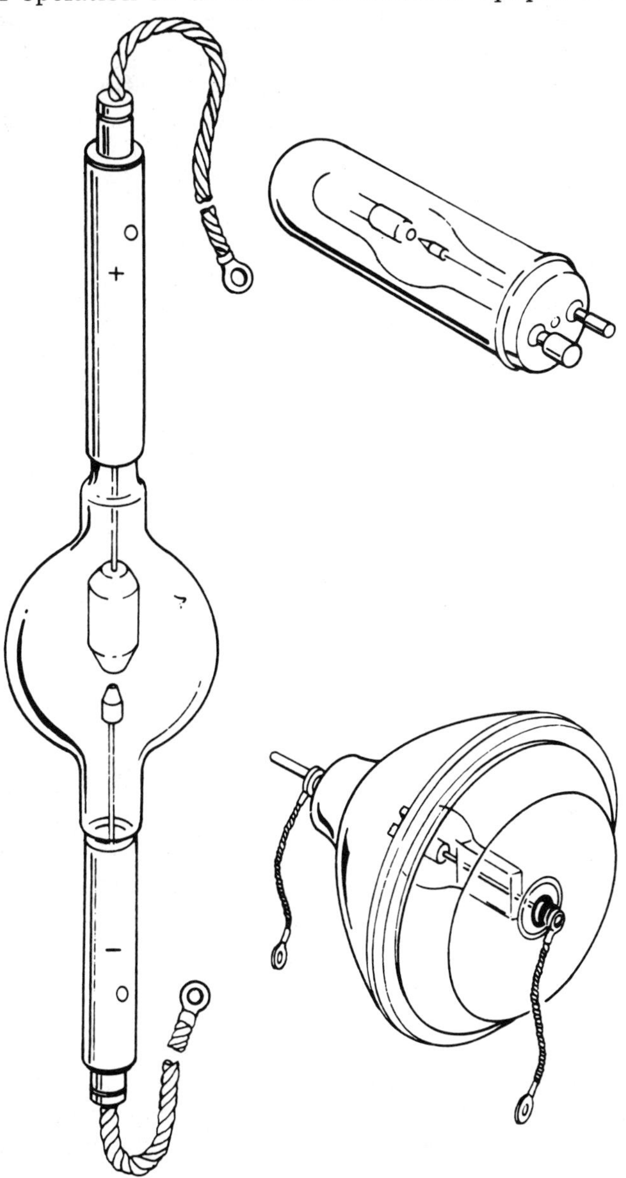

Figure 16-8 Xenon compact arc lamps.

500-W lamps has an ellipsoidal cool-beam reflector, designed to deliver maximum light with minimum infrared into special optical imaging systems. The 500-W lamp with tubular jacket is used in a number of optical devices, including one for the inspection of plate glass. The 500-W lamp is used in special optical devices, including solar radiation simulation for space research.

Sodium lab-arc. The NA-1 lamp is a small 500-lm source of low-pressure sodium radiation. Special laboratory equipment is available, including lamp and transformer. Principal application of the lamp is to generate modest concentrations of energy at the wavelength of the sodium resonance doublet, and 589 nm, for optical and other research purposes requiring essentially monochromatic light.

Tungsten arc. The 30A/PS22 lamp (Fig. 16-9) contains a tungsten arc with argon starting gas. The arc is used to heat a small, button-shaped tungsten electrode to incandescence, providing a sharply defined disc or relatively uniform luminance, and a color temperature of about 3000°K. It is used in a number of applications, which include photomicrography and optical comparators.

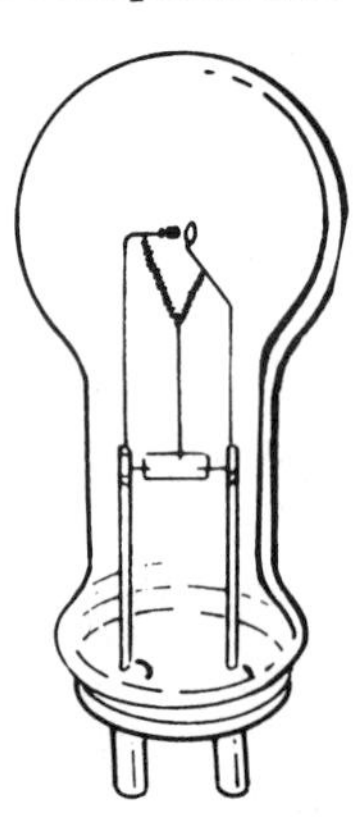

Figure 16-9 A typical tungsten-arc lamp.

Germicidal lamps. Germicidal lamps are actually low-pressure mercury lamps similar to fluorescent lamps in electrical characteristics, but made with special glass designed to pass for ultraviolet energy.

LUMINAIRE DATA

A luminaire is a complete lighting unit: lamp, sockets, and equipment for controlling light distribution and comfort. For example, the lens controls high-angle luminance where a specific light distribution light pattern is desired; diffusers are used where general diffusion of light is desired; shielding—in the form of louvers, baffles, and reflectors—is used to reduce glare and excessive brightness; and reflectors can be used to direct light in useful directions.

A lighting fixture may be classified by its distribution of light, type of lamp used, or description. The distribution of light is based on the percentage of lumens emitted above and below the horizontal.

Direct (0% to 10%): Ninety to 100% down. This type is most efficient from the standpoint of getting the maximum amount of light from the source to the working plane. On the other hand, this type may produce the greatest luminance differences between ceiling and luminaire, and produce the most shadows and glare.

Semi-direct (1% to 40% up): Sixty to 90% down. Most of the light is still down, but some is directed up to the ceiling.

General diffuse (40% to 60% down): Forty to 60% up. This type makes light available about equally in all directions. A modification of the general diffuse is the direct-indirect fixture, which is shielded to emit little light in the zone near the horizontal.

Semi-indirect (60% to 90% up): Ten to 40% down. Here, the greater percentage of the light is directed toward the ceiling and upper walls. The ceiling should be of high reflectance in order to reflect the light.

Indirect (90% to 100% up): Zero to 10% down. Totally indirect reflectors direct all of the light up to the ceiling. Some types are slightly luminous to offset the luminance difference between the luminaire and the bright ceiling. Shadows are at a minimum, although glare may be present due to a bright ceiling. Inside-frosted lamps should be used instead of clear-glass incandescent lamps to prevent streaks and striations on the ceiling. Low-luminance fluorescent lamps are recommended.

The classification of fixture according to source will be any of the three described previously—incandescent, fluorescent, or high intensity discharge lighting. Applications will include industrial, commercial, institutional, residential, and special-purpose applications such as for use in duct-tight, vapor-tight, and explosionproof areas.

Examples of several types of lighting fixtures are shown in Fig. 16-10. Note that light distribution curves, maximum spacing, and coefficients of utilization are given. These values are needed for lighting calculations and for selection of design and layout of lighting systems for areas of all types.

SPACING RATIO = 1.1

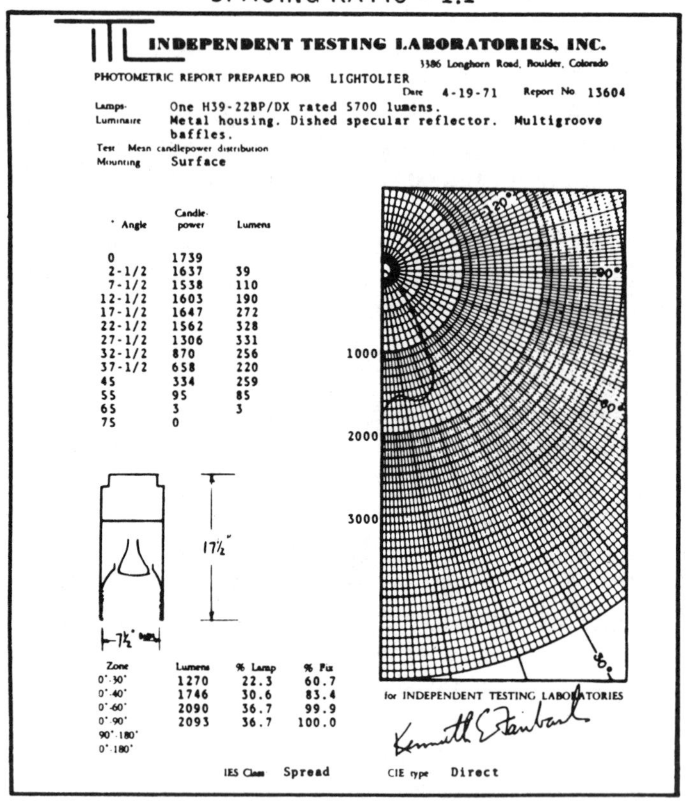

ITL INDEPENDENT TESTING LABORATORIES, INC.
3386 Longhorn Road, Boulder, Colorado

PHOTOMETRIC REPORT PREPARED FOR LIGHTOLIER

Date 4-19-71 Report No 13604

Lamps: One H39-22BP/DX rated 5700 lumens.
Luminaire: Metal housing. Dished specular reflector. Multigroove baffles.
Test: Mean candlepower distribution
Mounting: Surface

Angle	Candlepower	Lumens
0	1739	
2-1/2	1637	39
7-1/2	1538	110
12-1/2	1603	190
17-1/2	1647	272
22-1/2	1562	328
27-1/2	1306	331
32-1/2	870	256
37-1/2	658	220
45	334	259
55	95	85
65	3	3
75	0	

Zone	Lumens	% Lamp	% Fix
0°-30°	1270	22.3	60.7
0°-40°	1746	30.6	83.4
0°-60°	2090	36.7	99.9
0°-90°	2093	36.7	100.0
90°-180°			
0°-180°			

for INDEPENDENT TESTING LABORATORIES

IES Class Spread CIE type Direct

I.T.L. Report No. 13604C. These coefficients were computed by the Zonal-Cavity Method, I.E.S. Recommended Practice, and prepared from the candlepower distribution data given in Independent Testing Laboratories Report No. 13604, dated 4/19/71, and are based on a 20% Floor Cavity Reflectances.

	Ceiling Cavity Reflectance															
	80			70			50			30			10			0
	Wall Reflectance															
	50	30	10	50	30	10	50	30	10	50	30	10	50	30	10	0
Room Cavity Ratio	Coefficients of Utilization															
1	41	.40	.39	.40	.39	.38	.38	.38	.37	.37	.36	.36	.36	.35	.35	.34
2	.38	.36	.35	.37	.36	.35	.36	.35	.34	.35	.34	.33	.34	.33	.32	32
3	.35	.33	.32	.35	.33	.31	.34	.32	.31	.33	.31	.30	.32	.31	.30	.29
4	.33	.30	.29	.32	.30	.29	.31	.30	.28	.31	.29	.28	.30	.29	.28	.27
5	.30	.28	.26	.30	.28	.26	.29	.27	.26	.28	.27	.26	.28	.27	.25	.25
6	.28	.26	.24	.28	.26	.24	.27	.25	.24	.27	.25	.24	.26	.25	.24	.23
7	.26	.24	.22	.26	.24	.22	.25	.23	.22	.25	.23	.22	25	.23	.22	.21
8	.24	.22	.20	.24	.22	.20	.23	.21	.20	.23	.21	.20	.23	.21	.20	.19
9	22	20	.18	.22	.20	.18	.22	.20	.18	.21	.19	.18	.21	.19	.18	.17
10	20	18	.17	.20	.18	17	.20	.18	.17	.20	.18	.16	.19	.18	.16	.16

Figure 16-10 Several types of lighting fixtures.

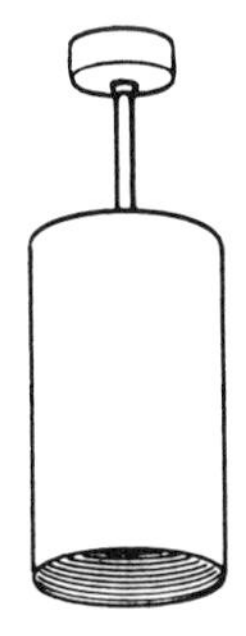

SPACING RATIO - 1.1

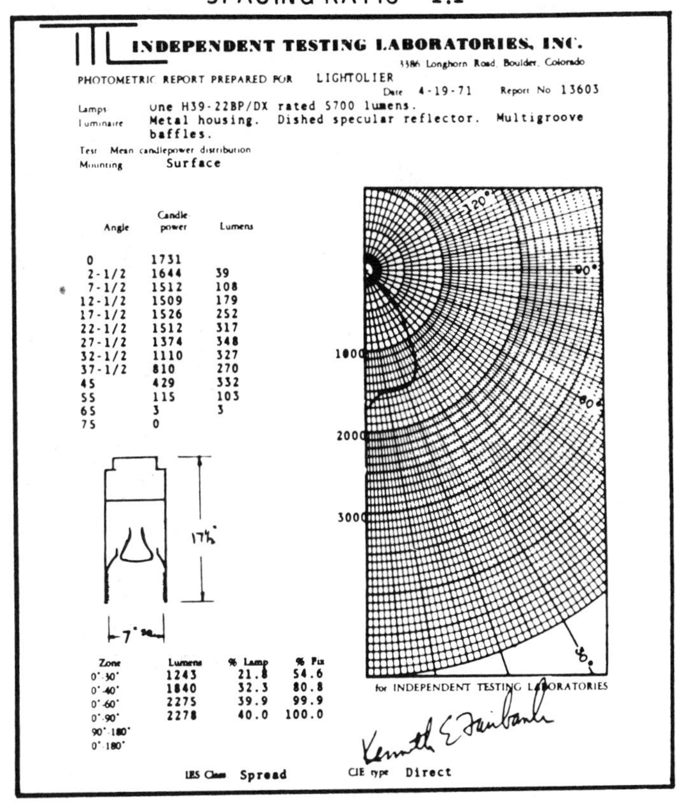

ITL INDEPENDENT TESTING LABORATORIES, INC.
3386 Longhorn Road, Boulder, Colorado

PHOTOMETRIC REPORT PREPARED FOR LIGHTOLIER
Date 4-19-71 Report No 13603

Lamps: One H39-22BP/DX rated 5700 lumens.
Luminaire: Metal housing. Dished specular reflector. Multigroove baffles.
Test: Mean candlepower distribution
Mounting: Surface

Angle	Candle-power	Lumens
0	1731	
2-1/2	1644	39
7-1/2	1512	108
12-1/2	1509	179
17-1/2	1526	252
22-1/2	1512	317
27-1/2	1374	348
32-1/2	1110	327
37-1/2	810	270
45	429	332
55	115	103
65	3	3
75	0	

Zone	Lumens	% Lamp	% Fix
0°-30°	1243	21.8	54.6
0°-40°	1840	32.3	80.8
0°-60°	2275	39.9	99.9
0°-90°	2278	40.0	100.0
90°-180°			
0°-180°			

for INDEPENDENT TESTING LABORATORIES

Kenneth E Fairbanks

IES Class Spread CIE type Direct

I.T.L. Report No. 13603C. These coefficients were computed by the Zonal-Cavity Method, I.E.S. Recommended Practice, and prepared from the candlepower distribution data given in Independent Testing Laboratories Report No. 13603, dated 4/19/71, and are based on a 20% Floor Cavity Reflectances.

	Ceiling Cavity Reflectance															
	80			70			50			30			10			0
	Wall Reflectance															
	50	30	10	50	30	10	50	30	10	50	30	10	50	30	10	0
Room Cavity Ratio	Coefficients of Utilization															
1	.44	.43	.42	.43	.42	.41	.42	.41	.40	.40	.40	.39	.39	.38	.38	.37
2	.41	.39	.38	.40	.39	.37	.39	.38	.36	.38	.37	.36	.37	.36	.35	.34
3	.38	.36	.34	.37	.35	.34	.36	.35	.33	.35	.34	.33	.34	.33	.32	.32
4	.35	.33	.31	.34	.32	.31	.34	.32	.30	.33	.31	.30	.32	.31	.29	.29
5	.32	.30	.28	.32	.29	.28	.31	.29	.27	.30	.29	.27	.30	.28	.27	.26
6	.30	.27	.25	.30	.27	.25	.29	.27	.25	.28	.26	.25	.28	.26	.25	.24
7	.28	.25	.23	.27	.25	.23	.27	.25	.23	.26	.24	.23	.26	.24	.23	.22
8	.25	.23	.21	.25	.23	.21	.25	.22	.21	.24	.22	.21	.24	.22	.21	.20
9	.23	.21	.19	.23	.20	.19	.23	.20	.19	.22	.20	.19	.22	.20	.18	.18
10	.21	.19	.17	.21	.19	.17	.21	.18	.17	.20	.18	.17	.20	.18	.17	.16

Figure 16-10 (cont.)

SPACING RATIO = 1.0

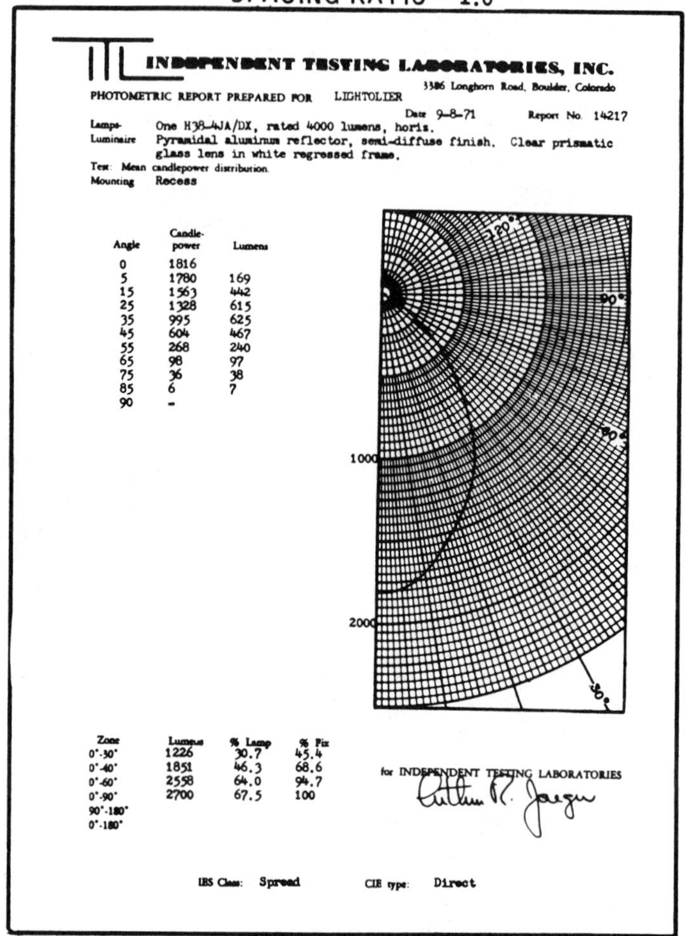

ITL INDEPENDENT TESTING LABORATORIES, INC.
3386 Longhorn Road, Boulder, Colorado

PHOTOMETRIC REPORT PREPARED FOR LIGHTOLIER

Date 9-8-71 Report No. 14217

Lamps One H38-4JA/DX, rated 4000 lumens, horiz.
Luminaire Pyramidal aluminum reflector, semi-diffuse finish. Clear prismatic glass lens in white regressed frame.
Test: Mean candlepower distribution.
Mounting Recess

Angle	Candle-power	Lumens
0	1816	
5	1780	169
15	1563	442
25	1328	615
35	995	625
45	604	467
55	268	240
65	98	97
75	36	38
85	6	7
90	-	

120°
90°
60°
1000
2000
30°

Zone	Lumens	% Lamp	% Fix
0°-30°	1226	30.7	45.4
0°-40°	1851	46.3	68.6
0°-60°	2558	64.0	94.7
0°-90°	2700	67.5	100
90°-180°			
0°-180°			

for INDEPENDENT TESTING LABORATORIES

IES Class: Spread CIE type: Direct

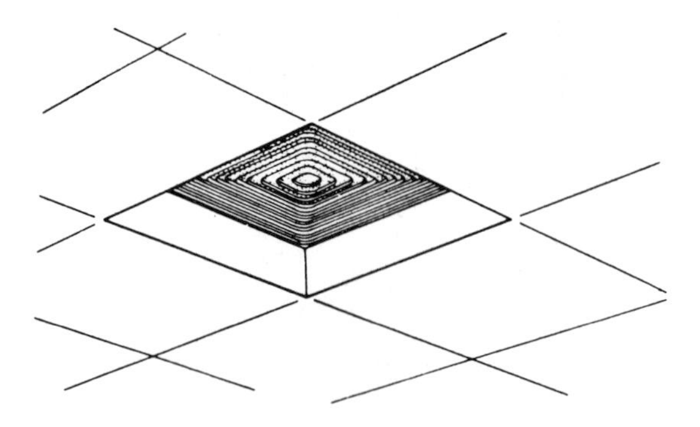

Figure 16-10 (cont.)

ILLUMINATION CALCULATION METHODS

Zonal-Cavity Method

The zonal-cavity method is the latest recommended method for determining the average maintained illumination level on the work plane in a given lighting installation, and also for determining the number of lighting fixtures required in a given area to provide a desired or recommended illumination level.

The Interior Illumination form in Fig. 16-11 is recommended

INTERIOR ILLUMINATION

FLOOR________ PROJECT________

ROOM________ FT. C. REQ'D________

WIDTH________ X LENGTH________ = AREA (A)________ SQ. FT.

CEILING HT.________ MOUNTING HT.________ RCR________

REFLECTION FACTORS: CEILING________% WALLS________% FLOOR________%

COEF. OF UTILIZATION (CU)________ MAINTENANCE FACTOR (MF)________

FIXTURE TYPE________ LAMPS/FIX.________ WATTS/FIX.________

LAMP TYPE________ LUMENS/LAMP________

$$\text{TOTAL LUMENS} = \frac{\text{FC} \times \text{A}}{\text{CU} \times \text{MF}} = \_\_\_\_\_\_ = \_\_\_\_\_\_$$

$$\text{NO. FIXTURES} = \frac{\text{TOTAL LUMENS}}{\text{LUMENS/LAMP} \times \text{LAMPS/FIX.}} = \_\_\_\_\_\_ = \_\_\_\_\_\_$$

ACTUAL NO. FIXTURES DESIGNED________

$$\text{REVISED FC} = \frac{\text{TOTAL FIX.} \times \text{LAMPS/FIX.} \times \text{LUMENS/LAMP} \times \text{CU} \times \text{MF}}{\text{AREA}} = \_\_\_\_\_\_$$

$$\text{WATTS/SQ. FT.} = \frac{\text{NO. FIXTURES} \times \text{WATTS/FIX.}}{\text{AREA}} =$$

CEILING CAVITY H_{cc}

LUMENAIRE PLANE

ROOM CAVITY H_{rc}

WORK PLAIN

FLOOR CAVITY H_{fc}

Figure 16-11 Interior illumination form.

when the zonal-cavity method is used. The illustration at the bottom of this form shows that a room or area is separated into three areas: (1) ceiling cavity, (2) room cavity, and (3) floor cavity. The cavity ratios are determined as follows:

$$\text{Room-Cavity Ratio (RCR):}\quad \text{RCR} = \frac{5h_{rc}(L + W)}{L + W}$$

$$\text{Ceiling-Cavity Ratio (CCR):}\quad \text{CCR} = \frac{5h_{cc}(L + W)}{L + W}$$

$$\text{Floor-Cavity Ratio (FCR):}\quad \text{FCR} = \frac{5h_{fc}(L + W)}{L + W}$$

The cavity heights are represented by h_{rc}, h_{cc}, and h_{fc} and are shown in the illustration. L is the room length, and W is the room width.

In using the form, record the room width and length, the area of the room, the ceiling height, and the mounting height of the lighting fixtures above the floor as well as the data required in the form.

After this information is inserted, calculate the cavity ratios as described previously and insert the resulting data in the space designated RCR.

Now select the effective ceiling reflectance (p_{cc}) from Fig. 16-12 for combination of ceiling and wall reflectances. Note that for a surface-mounted or recessed lighting fixture, CCR-0 and the ceiling reflectance may be used as the effective cavity reflectance.

The next step is to select the effective floor-cavity reflectance (p_{fc}) for the combination of floor and wall reflectances from Fig. 16-12.

Next the coefficient of utilization is determined by referring to a *coefficient of utilization* table for the lighting fixture under consideration. This figure is entered in the proper space on the form (CU). The maintenance factor (MF) is determined by the estimated amount of dirt accumulation on the fixtures prior to cleaning. This figure can vary and it takes some experience to select the right one. However, the following will act as a guide:

Very clean surroundings (hospitals, etc.)	0.80
Clean surroundings (restaurants, etc.)	0.75
Average surroundings (schools, etc.)	0.70
Below-average surroundings:	0.65
Dirty surroundings:	0.55

EFFECTIVE CEILING CAVITY REFLECTANCE (%)	80			70			50			10		
WALL REFLECTANCE (%)	50	30	10	50	30	10	50	30	10	50	30	10
ROOM CAVITY RATIO												
1	1.08	1.08	1.07	1.07	1.06	1.06	1.05	1.04	1.04	1.01	1.01	1.01
2	1.07	1.06	1.05	1.06	1.05	1.04	1.04	1.03	1.03	1.01	1.01	1.01
3	1.05	1.04	1.03	1.05	1.04	1.03	1.03	1.03	1.02	1.01	1.01	1.01
4	1.05	1.03	1.02	1.04	1.03	1.02	1.03	1.02	1.02	1.01	1.01	1.00
5	1.04	1.03	1.02	1.03	1.02	1.02	1.02	1.02	1.01	1.01	1.01	1.00
6	1.03	1.02	1.01	1.03	1.02	1.01	1.02	1.02	1.01	1.01	1.01	1.00
7	1.03	1.02	1.01	1.03	1.02	1.01	1.02	1.01	1.01	1.01	1.01	1.00
8	1.03	1.02	1.01	1.02	1.02	1.01	1.02	1.01	1.01	1.01	1.01	1.00
9	1.02	1.01	1.01	1.02	1.01	1.01	1.02	1.01	1.01	1.01	1.01	1.00
10	1.02	1.01	1.01	1.02	1.01	1.01	1.02	1.10	1.01	1.01	1.01	1.00

Figure 16-12 Effective floor-cavity reflectance table.

The manufacturer's catalog number is entered in the space marked Lamp Type, the number of lamps used in this fixture is entered next, and the total watts per fixture is next. Lamp lumens can be found in lamp manufacturers' catalogs under the type of lamp used in the fixture. The remaining calculations on the form are obvious.

Lighting fixture locations depend on the general architectural style, size of bays, type of lighting fixtures under consideration, and similar factors. However, in order to provide even distribution of illumination for an area, the permissible maximum spacing recommendations should not be exceeded. These recommendation ratios are usually supplied by the fixture manufacturers in terms of maximum spacing to mounting height. The fixtures, in some cases, will have to be located closer together than these maximums, in order to obtain the required illumination levels.

Point-by-Point Method of Calculating Illumination

The method for determining the average illumination of a given area was just described. However, it is sometimes desirable to know what the illumination level will be from one or more lighting fixtures upon a specified point within the area.

The point-by-point method accurately computes the level of illumination, in footcandles, at any given point in a lighting installation. This is accomplished by summing up all the illumination contributions, except surface reflection, to that point from every fixture individually. Since reflection from walls, ceilings, floors, and so on are not taken into consideration by this method, it is especially useful for calculations dealing with very large areas, outdoor lighting, and areas where the room surfaces are dark or dirty. With the aid of a candlepower distribu-

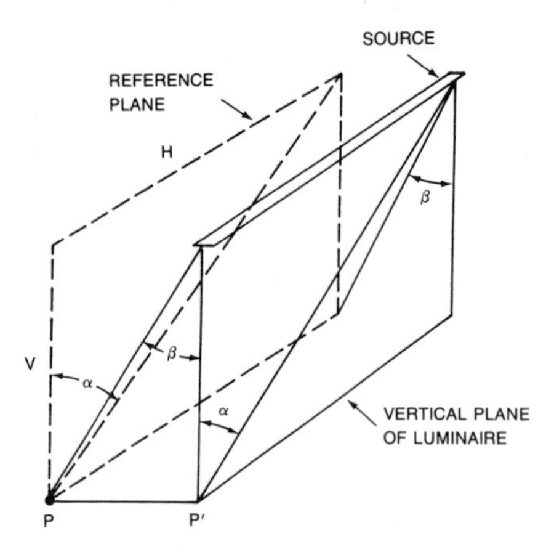

Figure 16-13 Example showing use of point-by-point illumination method.

tion curve, we may calculate footcandle (ft c) values for specific points as follows:

$$\text{ft c} = \text{candlepower} \times H/d^3$$

or

$$\text{candlepower} \times \cos^3 0/H^2$$

Vertical surfaces

$$\text{ft c} = \text{candlepower} \times R/d^3$$

or

$$\text{candlepower} \times \cos^3 0 \times \sin 0/h^2$$

In using the point-by-point method, a specific point is selected at which it is desired to know the illumination level, at, for example, point

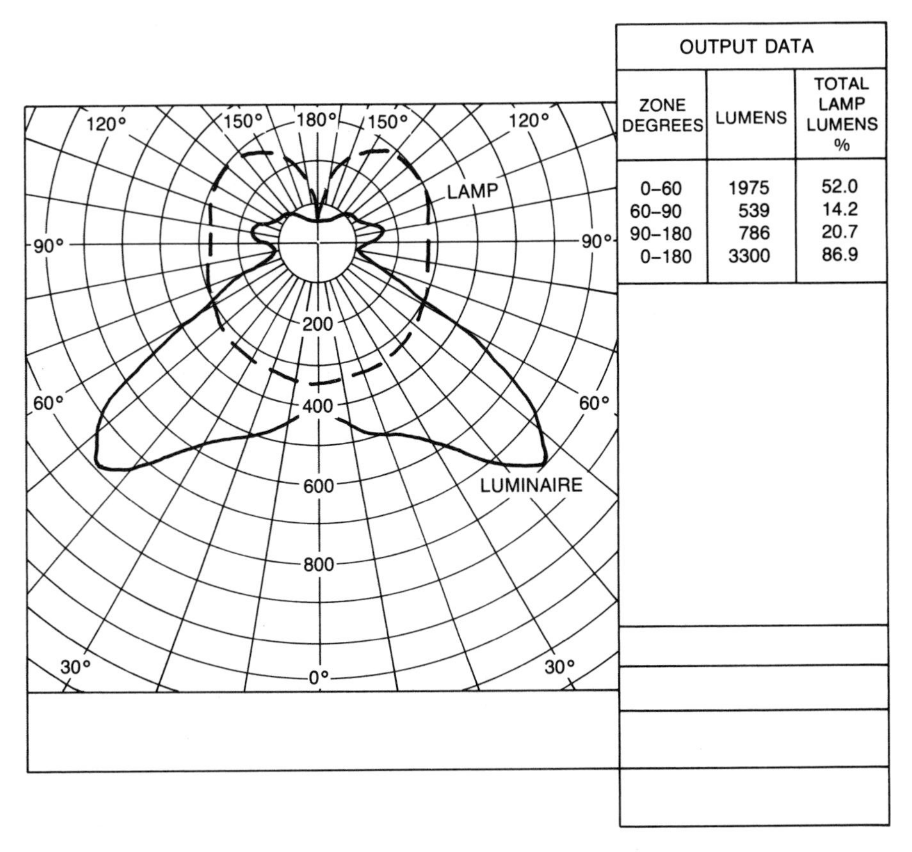

ZONE DEGREES	LUMENS	TOTAL LAMP LUMENS %
0–60	1975	52.0
60–90	539	14.2
90–180	786	20.7
0–180	3300	86.9

Figure 16-14 A typical candlepower distribution curve.

P in Fig. 16-13. Once the seeing task or point has been determined, the illumination level at the point can be calculated.

It is obvious that the illumination at point *P*, or at any point in the area, is due to light coming from all of the lighting fixtures. In this case, the calculations must be repeated to determine the amount of

$\theta°$	$\sin\theta$	$\cos\theta$	$\cos^2\theta$	$\cos^3\theta$	$\tan\theta$	$\theta°$	$\sin\theta$	$\cos\theta$	$\cos^2\theta$	$\cos^3\theta$	$\tan\theta$
0	0.0	1.000	1.000	1.000	0.0	46	0.719	0.695	0.483	0.335	1.035
1	.0175	1.000	1.000	1.000	.0174	47	.731	.682	.465	.317	1.072
2	.0349	0.999	0.999	.998	.0349	48	.743	.669	.448	.300	1.110
3	.0523	.999	.997	.996	.0524	49	.755	.656	.430	.282	1.150
4	.0698	.998	.995	.993	.0699	50	.766	.643	.413	.266	1 .191
5	.0872	.996	.992	.989	.0874	51	.777	.629	.396	.249	1.234
6	.105	.885	.989	.984	.1051	52	.788	.616	.379	.233	1.279
7	.122	.993	.985	.978	.1227	53	.799	.602	.362	.218	1.327
8	.138	.990	.981	.971	.1405	54	.809	.588	.345	.203	1.376
9	.156	.988	.976	.964	.1583	55	.819	.574	.329	.189	1.428
10	.164	,875	.970	.955	.1763	56	.829	.559	.313	.15	1.482
11	.191	.982	.964	.946	.1943	57	.839	.545	.297	.162	1.539
12	.208	.978	.957	.936	.2125	58	.848	.530	.281	.149	1.600
13	.225	.974	.949	.925	.2308	59	.857	.515	.265	.137	1.664
14	.242	.970	.941	.913	.2493	60	.866	.500	.250	.125	1.732
15	.259	.966	.933	.901	.2679	61	.875	.485	.235	.114	1.804
16	.276	.961	.924	.888	.2867	62	.883	.470	.220	.103	1.880
17	.292	.956	.915	.875	.3057	63	.891	.454	.206	.0936	1.962
18	.309	.951	.905	.860	.3249	64	.899	.438	.192	.0842	2.050
19	.326	.846	.894	.845	.3443	65	.906	.423	.179	.0755	2.144
20	.342	.940	.883	.830	.3639	66	.914	.407	.165	.0673	2.246
21	.358	.934	.872	.814	.3838	67	.921	.391	.153	.0597	2.355
22	.375	.927	.860	.797	.4040	68	.927	.375	.140	.0526	2.475
23	.391	.921	.847	.780	.4244	69	.934	.358	.128	.0460	2.605
24	.407	.914	.835	.762	.4452	70	.940	.342	.117	.0400	2.747
25	.423	.906	.821	.744	.4663	71	.946	.326	.106	.0347	2.904
26	.438	.899	.808	.726	.4877	72	.951	.309	.0955	.0295	3.077
27	.454	.891	.704	.707	.5095	73	.956	.292	.0855	.0250	3.270
28	.470	.883	.780	.688	.5317	74	.961	.276	.0762	.0211	3.487
29	.475	.875	.765	.669	.5543	75	.966	.259	.0670	.0173	3.732
30	.500	.766	.659	.659	.5773	76	.970	.242	.0585	.0142	4.010
31	.515	.857	.735	.630	.6008	77	.974	.225	.0506	.0114	4.331
32	.530	.848	.719	.610	.6248	78	.978	.208	.0432	.0090	4.704
33	.545	.839	.703	.590	.6494	79	.982	.191	.0364	.0070	5.144
34	.559	.829	.687	.570	.6745	80	.985	.174	.0302	.0052	5.671
35	.564	.819	.671	.550	.7002	81	.988	.156	.0245	.0038	6.313
36	.588	.809	.655	.530	.7265	82	.990	.139	.0194	.0027	7.115
37	.692	.799	.638	.509	.7535	83	.993	.122	.0149	.0018	8.144
38	.616	.788	.621	.489	.7812	84	.995	.105	.0109	.0011	9.514
39	.629	.777	.604	.469	.8097	85	.996	.0872	.0076	.0007	11.430
40	.643	.766	.576	.459	.7381	86	.8876	.0698	.0048	.0003	14.300
41	.656	.755	.570	.430	.8692	87	.9986	.0523	.0027	.0001	19.080
42	.669	.743	.552	.410	.9004	88	.9993	.0349	.0012	.0000	28.630
43	.682	.731	.535	.381	.8325	89	.9998	.0175	.0003	.0000	57.280
44	.695	.719	.517	.372	.9656	90	1.0000	0.0000	.0000	.0000	Infinite
45	.707	.707	.500	.354	1.0000						

Figure 16-15 Table of trigonometric functions.

light that each fixture contributes to the point; the total amount is the sum of all the contributing values.

Before attempting any actual calculations using the point-by-point method, a knowledge of candlepower distribution curves and a review of trigonometric functions is necessary.

A candlepower distribution curve or graph consists of lines plotted on a polar diagram, which show graphically the distribution of the light flux in some given plane around the actual light source. It also shows the apparent candlepower intensities in various directions about the light source. Figure 16-14 illustrates a typical candlepower distribution curve.

A table of trigonometric functions (Fig. 16-15) will be helpful in determining the degrees of the angle from the light source to the point in question. This is necessary to pick off the candlepower from the photometric distribution curve and also to use in the equations.

OUTLET LOCATION

There is no general rule for locating lighting outlets for general illumination, but it is usually desirable to locate lighting fixtures so that the illumination in a given area is uniform. Where the number and location of lighting outlets cannot be determined from the architectural or electrical drawings or by the decor layout, it usually is desirable to arrange the lighting outlets in the form of squares or rectangles (see Fig. 16-16).

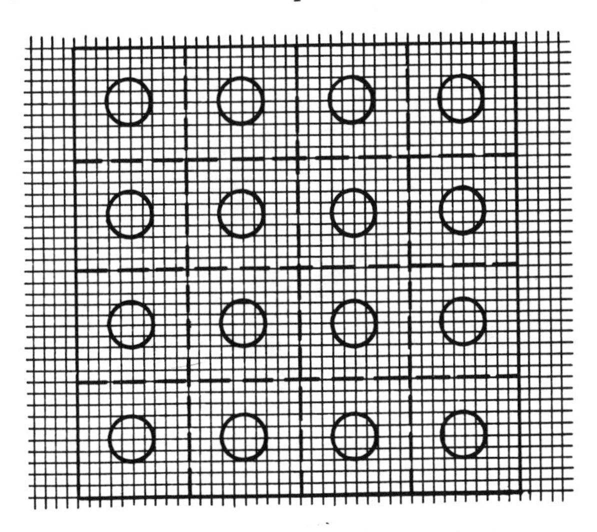

Figure 16-16 Recommended lighting outlet layout.

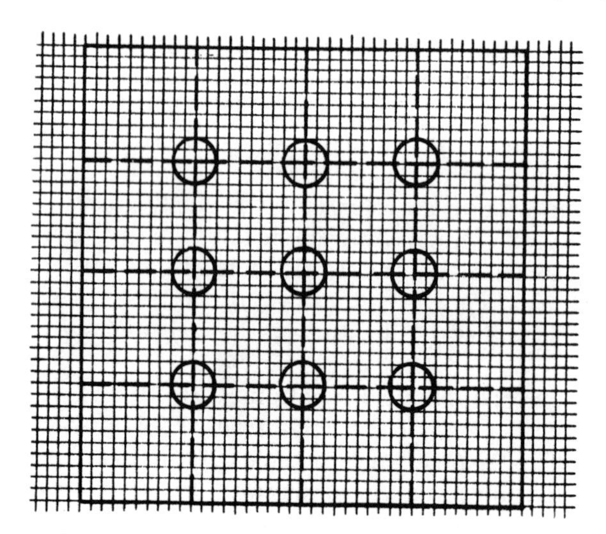

Figure 16-17 Inefficient lighting outlet layout.

In doing so, the outlets should be placed at the centers of the squares and not at the corners as shown in Fig. 16-17. The latter arrangement will provide most of the light near the center of the room and will therefore create a very low intensity of illumination near the walls.

17

Wiring in Hazardous Locations

Wiring in hazardous locations was briefly covered in Chapter 5, Electrical Boxes. However, additional information is in order.

Articles 500 through 503 of the NE Code cover the requirements of electrical equipment and wiring for all voltages in locations where fire or explosion hazards may exist. Locations are classified depending on the properties of the flammable vapors, liquids, or gases, or combustible dusts or fibers that may be present, as well as the likelihood that a flammable or combustible concentration or quality is present.

Any area in which the atmosphere or a material in the area is such that the arcing of operating electrical contacts, components, and equipment may cause an explosion or fire is considered a hazardous location. In all such cases, explosionproof equipment, raceways, and fittings are used to provide an explosionproof wiring system.

Hazardous locations have been classified in the NE code into certain class locations. Various atmospheric groups have been established on the basis of the explosive character of the atmosphere for the testing and approval of equipment for use in the various groups.

GROUPS

Groups A, B, C, and D (Class I)

Combustible and flammable gases and vapors are divided into four groups, the classification involving determinations of maximum explosion pressures, maximum safe clearance between parts of a clamped joint in an enclosure, and the minimum ignition temperature of the atmospheric mixture.

In general, the hazardous properties of the substances are greater for Group A. Group B is the next most hazardous, then Group C, with Group D the least hazardous. However, all four groups are extremely dangerous. Equipment to be used in these atmospheres must not only be approved for Class I, but also for the specific group of gases or vapors that will be present.

Groups E, F, and G (Class II)

Combustible dusts are divided into these three groups, the classification involving determinations of the tightness of the joints of assembly and shaft openings, the blanketing effect of layers of dust on the equipment that may cause over-heating, the electrical conductivity of the dust, and the ignition temperatures of the dust.

Group E Atmospheres

This group contains metal dust, including aluminum, magnesium, and their commercial alloys, and other metals of similarly hazardous characteristics having resistivity of 10^2 Ω-cm or less.

Group F Atmospheres

This group contains carbon black, charcoal, coal, or coke dusts, which have more than a certain percentage total volatile material or atmospheres containing these dusts sensitized by other materials, and having resistivity greater than 10^2 Ω-cm but equal to or less than 10^8 Ω-cm.

Group G Atmospheres

This group contains flour, starch, grain, or combustible plastics or chemical dusts having resistivity greater than 10^8 Ω-cm.

Equipment to be used in these atmospheres must not only be approved for Class II, but also for the specific group of dust that will be present.

DIVISIONS

Classes and groups are divided into two divisions: 1 and 2. Generally, Division 1 refers to classes and groups where ignitible mixtures exist under normal everyday conditions while Division 2 refers to classes and groups where ignitible mixtures exist only under unusual conditions.

EXPLOSIONPROOF EQUIPMENT

The wide assortment of explosionproof equipment now available makes it possible to provide adequate electrical installations under any hazardous condition likely to be involved on any industrial project. However, the electrical engineer and workers on the job must be thoroughly familiar with all the NE Code requirements, and know what fittings and other explosionproof equipment are available; furthermore, they must be knowledgeable about their installation.

There is a rather common misconception that explosionproof equipment is gastight. It would be inadvisable to make an entire wiring system gastight. Whenever an enclosure would be opened for servicing, the explosive mixture could enter and be trapped in the enclosure. This trapped atmosphere would explode the instant the apparatus was again operated. The explosion could develop sufficient pressure to burst a gastight enclosure and allow flames to escape into the surrounding atmosphere.

The requirement, therefore, is not that enclosures be gastight, but that they be designed and manufactured strong enough to contain an explosion and prevent the escape of flame or heat that could ignite surrounding atmospheres. Burned gases do escape from explosionproof equipment, but their escape path has been engineered so the temperature of the gas is well below the ignition point when it releases into the surrounding atmosphere. Most explosionproof fittings are designed to withstand a hydrostatic test of four times the maximum internal explosion pressure that could be developed from a gas explosion.

CHOOSING EQUIPMENT

To accurately determine the type of equipment needed for a given location, it is essential to classify the hazards according to the NE Code, Article 500. As stated before, explosive atmospheres are divided into Class I, Groups A, B, C, and D; Class II, Groups E, F, and G; and Class III, according to the characteristics of the gas, vapor, or dust involved.

Devices qualified for use in Class I locations are not necessarily effective for classes II or III. Some are suitable and are so indicated. Conceivably a device appropriate for a Class I location might become covered with dust and overheat in a Class II location. Most products designed for use in classified locations are prominently indicated by class and group on each listing page in the manufacturer's general catalog.

SEALING FITTINGS

Sealing fittings prevent the passage of gases, vapors, or flames from one portion of a conduit system to another. They also restrict the accumulation of large amounts of ignitible gases or vapors and thereby limit explosive pressure.

Many electrical designers use conventional wiring symbols on the working drawings and merely note the class of hazard location, and that all wiring therein must be installed according to the NE Code. However, to ensure adequate protection, the working drawings should indicate the location of each required seal-off. The better drawings even indicate the type; that is, EYS, SF and BYD.

In each conduit run leaving the Class I, Division 1 hazardous area, the sealing fitting may be located on either side of the boundary of the

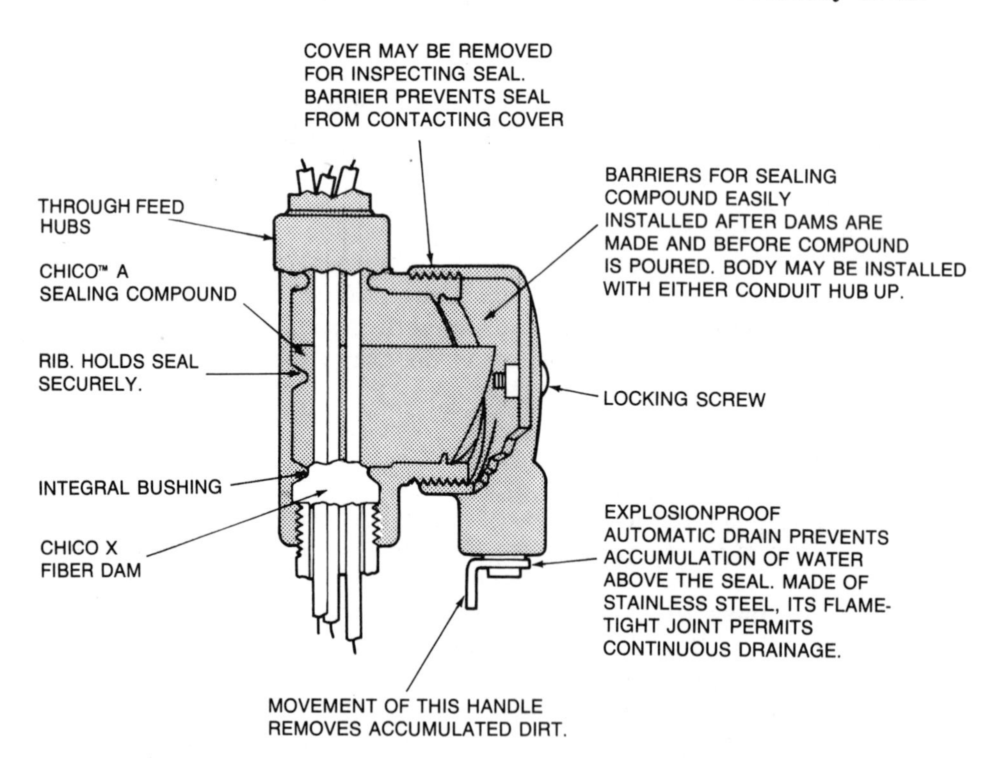

Figure 17-1 Example of seal-off application. (Courtesy of Crouse-Hinds.)

hazardous area, but it must be designed and installed so that any gases or vapors that may enter the conduit system, within the Division 1 hazardous area, will not enter or be communicated to the conduit beyond the seal. There can't be any union, coupling, box, or fitting in the conduit between the sealing fitting and the point at which the conduit leaves the Division 1 hazardous area. See Fig. 17-1.

Sealing compound must be approved for the purpose, must not be affected by the surrounding atmosphere or liquids, and must not have a melting point of less than 200° F. Most sealing-compound kits contain a powder in a polyethylene bag within an outer container. To mix, remove the bag of powder, fill the outside container, and pour in the powder and mix.

To pack the seal-off, remove the threaded plug or plugs from the fitting and insert the asbestos fiber supplied with the packing kit. Tamp the fiber between the wires and the hub before pouring the sealing compound into the fitting. Then pour in the sealing cement and reset the threaded plug tightly. The fiber packing prevents the sealing compound (in the liquid state) from entering the conduit lines.

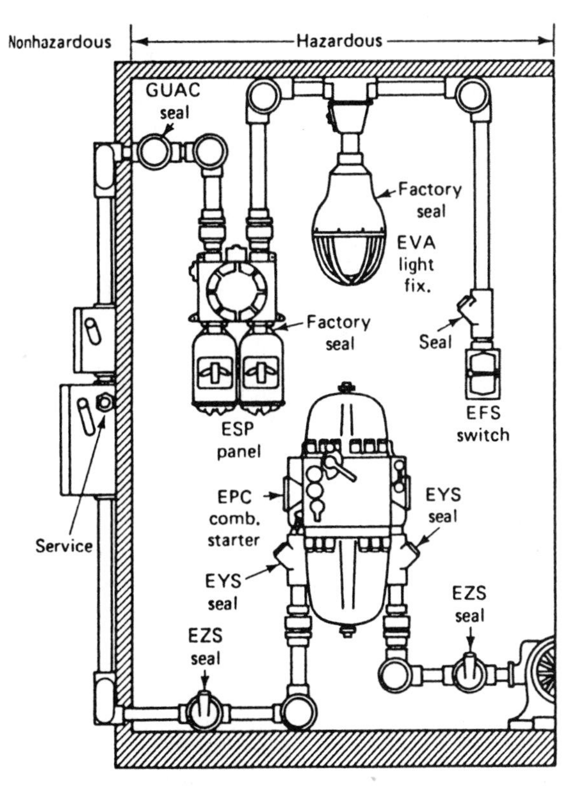

Figure 17-2 Typical seal-off fittings. (Courtesy of Crouse-Hinds.)

The seal-off fittings in Fig. 17-2 are typical of those used. One type is used for vertical mounting and is provided with a threaded, plugged opening into which the sealing cement is poured. The other has an additional plugged opening in the lower hub to facilitate packing fiber around the conductors in order to form a dam for the sealing cement.

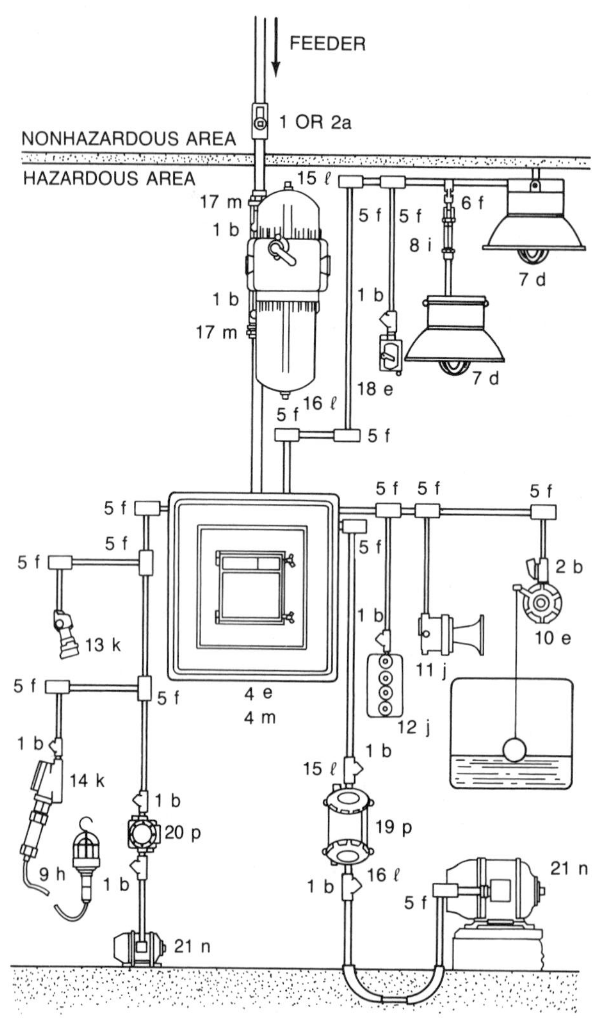

Key to Numerals

1. Sealing fitting. EYS for horizontal or vertical.
2. Sealing fitting. EZS for vertical or horizontal conduits.
3. Circuit breaker, Type EPC.
4. Panelboard, D2BP, N2PB. Branch circuits are factory sealed.
5. Junction box or conduit fitting. NJB. Obround® .
6. Fixture hanger. AHG, GS, UNJ.
7. Lighting fixture. VMV (CHAMP™).
8. Flexible-fixture support.
9. Handlamp. EVH.
10. Float switch. EMS.
11. Signal ETH horns and sirens. ESR belts.
12. Visularm™ EKP.
13. Plug receptacle. CES delayed action.
14. Plug receptacle. FSQ. Interlocked with switch.
15. Breather. ECD.
16. Drain. ECD.
17. Union. UNY.
18. Switch, Series EFS.
19. Manual line starter. FLF.
20. Manual line starter. GUSC.
21. Motors Suitable for Class 1, Division 2 locations.

National Electrical Code References

a Sec. 501-5(b)(2). Seal required where conduit passes from hazardous to nonhazardous area.
b Sec. 501-5(b)(1). Seals required within 18 in. of all arcing devices.
c Sec. 384-16. Circuit breaker protection required ahead of panelboard.
d Sec. 501-9(b)(2). All fixed lighting fixtures shall be enclosed and gasketed and not exceed ignition temperature of the gas.
e Sec. 501-6(b)(1). Most arcing devices must be explosion proof.
f Sec. 501-4(b). All boxes must be threaded for rigid or IMC conduit.
g Sec. 501-9(b)(1). All portable lighting fixtures must be explosion proof.
h Sec. 501-9(b)(3). Pendant fixture stems must be threaded rigid conduit or IMC. Rigid stems if over 12 in. must have flexible connector, or must be braced.
i Sec. 501-14(b). All signaling equipment must be approved for class I location.
j Sec. 501-12. Receptacles and plugs must be explosion proof and provide grounding connections for portable equipment.
k Sec. 501-5(f). Breathers and drains needed in all humid locations.
ℓ Sec. 501-4(b). Not all joints and fittings are required to be explosion proof.
m Sec. 501-8(b). Motor shall be suitable for Division 2.
n Art. 430. Motor overcurrent protection.

Figure 17-3 A practical representation of explosionproof fittings. (Courtesy of Crouse–Hinds.)

Most other explosionproof fittings are provided with threaded hubs for securing the conduit as described previously. Typical fittings include switch and junction boxes, conduit bodies, union and connectors, flexible couplings, explosionproof lighting fixtures, receptacles, and panelboard and motor starter enclosures. A practical representation of these and other fittings is shown in Fig. 17-3.

The NE Code recommends that whenever possible electrical equipment for hazardous locations be located in less hazardous areas. It also suggests that by adequate, positive-pressure ventilation from a clean source of outside air, the hazards may be reduced or hazardous locations limited or eliminated. In many cases the installation of dust-collecting systems can greatly reduce the hazards in a Class II area.

Glossary

Accessible (as applied to wiring methods): Capable of being removed or exposed without damaging the building structure or finish, or not permanently closed in by the structure or finish of the building.

Accessible (as applied to equipment): Admitting close approach—not guarded by locked doors, elevation, or other effective means.

Aggregate: Inert material mixed with cement and water to produce concrete.

Ampacity: Current-carrying capacity expressed in amperes.

Appliance: Utilization equipment, generally equipment other than industrial, normally built in standardized sizes or types and installed or connected as a unit to perform one or more functions, such as clothes washing, air conditioning, food mixing, deep frying.

Appliance, fixed: An appliance that is fastened or otherwise secured at a specific location.

Appliance, portable: An appliance that is actually moved or can easily be moved from one place to another in normal use.

Appliance, stationary: An appliance that is not easily moved from one place to another in normal use.

Approved: Acceptable to the authority enforcing the NE Code.

Attachment plug (plug cap or Cap): A device that, upon insertion in a receptacle, establishes a connection between the conductors of the attached flexible cord and the conductors connected permanently to the receptacle.

Automatic: Self-acting, operating by its own mechanism when actuated by some impersonal influence, such as a change in current strength, pressure, temperature, or mechanical configuration.

Backfill: Loose earth placed outside foundation walls for filling and grading.

Bearing plate: Steel plate placed under one end of a beam or truss for load distribution.

Bearing wall: Wall supporting a load other than its own weight.

Bench mark: Point of reference from which measurements are made.

Bonding jumper: A reliable conductor used to ensure the required electrical conductivity between metal parts required to be electrically connected.

Branch circuit: That portion of a wiring system extending beyond the final overcurrent device protecting the circuit.

Branch circuit, appliance: A circuit supplying energy to one or more outlets to which appliances are to be connected; such circuits have no permanently connected lighting fixtures and are not a part of an appliance.

Branch circuit, general purpose: A branch circuit that supplies a number of outlets for lighting and appliances.

Branch circuit, individual: A branch circuit that supplies only one piece of utilization equipment.

Bridging: System of bracing between floor beams to distribute floor load.

Building: A structure that stands alone or that is cut off from adjoining structures by fire walls with all openings therein protected by approved fire doors.

Cabinet: An enclosure designed for either surface or flush mounting and provided with a frame, mat, or trim in which swinging doors are hung.

Cavity wall: Wall built of solid masonry units arranged to provide air space within the wall.

Chase: Recess in inner face of masonry wall providing space for pipes, ducts, or both.

Circuit breaker: A device designed to open and close a circuit by nonautomatic means and to open the circuit automatically on a predetermined overload of current, without injury to itself when properly applied within its rating.

Column: Vertical load-carrying member of a structural frame.

Concealed: Rendered inaccessible by the structure or finish of the building. Wires in concealed raceways are considered concealed, even though they may become accessible by withdrawing them.

Conductor, bare: A conductor without any covering or insulation.

Conductor, covered: A conductor having one or more layers of nonconducting materials that are not recognized as insulation under the NE Code.

Conductor, insulated: A conductor covered with material recognized by the NE Code as insulation.

Connector, pressure (solderless): A connector that establishes the connection between two or more conductors or between one or more conductors and a terminal by means of mechanical pressure and without the use of solder.

Continuous load: A load in which the maximum current is expected to continue for three hours or more.

Contour line: A line on a map denoting elevations, a line connecting points with the same elevation.

Controller: A device, or group of devices, that serves to govern, in some predetermined manner, the electric power delivered to the apparatus to which it is connected.

Cooking unit, counter mounted: An assembly of one or more domestic surface heating elements for cooking purposes, designed to be flush mounted in, or supported by, a counter and complete with internal wiring and inherent or separately mounted controls.

Crawl space: Shallow space between the first tier of beams and the ground (no basement).

Curtain wall: Nonbearing wall between piers or columns for the enclosure of the structure; not supported at each story.

Demand factor: In any system or part of a system, the ratio of the maximum demand of the system, or part of the system, to the total connected load of the system, or part of the system under consideration.

Disconnecting means: A device, or group of devices, or other means whereby the conductors of a circuit can be disconnected from their source of supply.

Double-strength glass: Sheet glass that is 1/8 in. thick (single-strength glass is 1/10 in. thick).

Dry wall: Interior wall construction consisting of plaster boards, wood paneling, or plywood nailed directly to the studs without application of plaster.

Duty, continuous: A requirement of service that demands operation at a substantially constant load for an indefinitely long time.

Duty, intermittent: A requirement of service that demands operation for alternate intervals of (1) load and no load, (2) load and rest, or (3) load, no load, and rest.

Duty, periodic: A type of intermittent duty in which the load conditions regularly recur.

Duty, short-time: A requirement of service that demands operations at loads and for intervals of time, both of which may be subject to wide variation.

Elevation: Drawing showing the projection of a building on a vertical plane.

Enclosed: Surrounded by a case that will prevent anyone from accidentally contacting live parts.

Equipment: A general term including material, fittings, devices, appliances, fixtures, apparatus, and the like used as a part of, or in connection with, an electrical installation.

Expansion bolt: Bolt with a casing arranged to wedge the bolt into a masonry wall to provide an anchorage.

Expansion joint: Joint between two adjoining concrete members arranged to permit expansion and contraction with changes in temperature.

Exposed (as applied to live parts): Live parts that a person could inadvertently touch or approach nearer than a safe distance. This term is applied to parts not suitably guarded, isolated, or insulated.

Exposed (as applied to wiring method): Not concealed.

Externally operable: Capable of being operated without exposing the operator to contact with live parts.

Facade: Main front of a building.

Feeder: The conductors between the service equipment, or the generator switchboard of an isolated plant, and the branch-circuit overcurrent device.

Fire stop: Incombustible filler material used to block interior draft spaces.

Fitting: An accessory such as a locknut, bushing, or other part of a wiring system that is primarily intended to perform a mechanical rather than an electrical function.

Flashing: Strips of sheet metal bent into an angle between the roof and wall to make a watertight joint.

Footing: Structural unit used to distribute loads to the bearing materials.

Frostline: Deepest level below grade to which frost penetrates in a geographic area.

Garage: A building or portion of a building in which one or more self-propelled vehicles carrying volatile, flammable liquid for fuel or power are kept for use, sale, storage, rental, or repair.

Isolated: Not readily accessible to persons unless special means for access are used.

Jamb: Upright member forming the side of a door or window opening.

Lally column: Compression member consisting of a steel pipe filled with concrete under pressure.

Laminated wood: Wood built up of plies or laminations that have been joined either with glue or with mechanical fasteners. Usually, the plies are too thick to be classified as veneer and the grain of all plies is parallel.

Lighting outlet: An outlet intended for the direct connection of a lampholder, lighting fixture, or pendant cord terminating in a lampholder.

Location, damp: A location subject to a moderate amount of moisture, such as some basements, some barns, some cold-storage warehouses, and the like.

Location, dry: A location not normally subject to dampness or wetness. A location classified as dry may be temporarily subject to dampness or wetness, as in the case of a building under construction.

Location, wet: A location subject to saturation with water or other liqquids, such as locations exposed to weather, washrooms in garages, and similar locations. Installations that are located underground or in concrete slabs, or masonry in direct contact with the earth are considered wet locations.

Low-energy power circuit: A circuit that is not a remote-control or signal circuit but whose power supply is limited in accordance with the requirements of Class-2 remote-control circuits.

Multioutlet assembly: A type of surface or flush raceway designed to hold conductors and attachment plug receptacles and assembled in the field or at the factory.

Nonautomatic: Used to describe an action requiring human intervention for its control.

Nonbearing wall: Wall that carries no load other than its own weight.

Outlet: In the wiring system, a point at which current is taken to supply utilization equipment.

Outline lighting: An arrangement of incandescent lamps or gaseous tubes to outline and call attention to certain features such as the shape of a building or the decoration of a window.

Oven, wall-mounted: A domestic oven for cooking purposes designed for mounting into or onto a wall or other surface.

Panelboard: A single panel or group of panel units designed for assembly in the form of a single panel; includes buses and may come with or without switches and/or automatic overcurrent protective devices for the control of light, heat, or power circuits of small individual as well as aggregate capacity. It is designed to be placed in a cabinet or cutout box placed in or against a wall or partition and accessible only from the front.

Pilaster: Flat square column attached to a wall and projecting about a fifth of its width from the face of the wall.

Plenum: Chamber or space forming a part of an air-conditioning system.

Precast concrete: Concrete units (such as piles or vaults) cast in a location separate from the construction site and set in place.

Qualified person: One familiar with the construction and operation of a given apparatus and the hazards involved.

Raceway: Any channel designed expressly for holding wires, cables, or bus bars and used solely for this purpose.

Rainproof: Constructed, protected, or treated as to prevent rain from interfering with successful operation of the apparatus.

Raintight: Constructed or protected so that exposure to a beating rain will not result in the entrance of water.

Readily accessible: Capable of being reached quickly, for operation, renewal, or inspections, without requiring those to whom ready access is requisite to climb over or remove obstacles or resort to portable ladders, chairs, and so on.

Receptacle (convenience outlet): Contact device installed at an outlet for the connection of an attachment plug.

Receptacle outlet: An outlet where one or more receptacles are installed.

Remote-control circuit: Any electrical circuit that controls any other circuit through a relay or an equivalent device.

Riser: Upright member of stair extending from tread to tread.

Roughing in: Installation of all concealed electrical wiring; includes all electrical work done before finishing.

Sealed (hermetic-type) motor compressor: A mechanical compressor consisting of a compressor and a motor, both of which are enclosed in the same sealed housing, with no external shaft or shaft seals, the motor operating in the refrigerant atmosphere.

Service: The conductors and equipment used for delivering energy from the electricity supply system to the wiring system of the premises served.

Service cable: The service conductors made up in the form of a cable.

Service conductors: The supply conductors that extend from the street main or transformers to the service equipment of the premises being supplied.

Service drop: The overhead service conductors from the last pole, or other aerial support, to and including the splices, if any, that connect to the service-entrance conductors at the building or other structure.

Service-entrance conductors, underground system: The service conductors between the terminals of the service equipment and the point of connection to the service lateral.

Service equipment: The necessary equipment, usually consisting of a circuit breaker, or switch and fuses, and their accessories, located near the point of entrance of supply conductors to a building and intended to constitute the main control and means of cutoff for the supply to that building.

Service lateral: The underground service conductors between the street main, including any risers at a pole or other structure or from transformers, and the first point of connection to the service-entrance conductors in a terminal box, meter, or other enclosure with adequate space, inside or outside the building wall. Where there is no terminal box, meter, or other enclosure with adequate space, the point of connection shall be considered to be the point of entrance of the service conductors into the building.

Service raceway: The rigid metal conduit, electrical metallic tubing, or other raceway that encloses the service-entrance conductors.

Setting (of circuit breaker): The value of the current at which the circuit breaker is set to trip.

Sheathing: First covering of boards or paneling nailed to the outside of the wood studs of a frame building.

Siding: Finishing material that is nailed to the sheathing of a wood frame building and that forms the exposed surface.

Signal circuit: Any electrical circuit supplying energy to an appliance that gives a recognizable signal.

Soffit: Underside of a stair, arch, or cornice.

Soleplate: Horizontal bottom member of wood-stud partition.

Studs: Vertically set skeleton members of a partition or wall to which lath is nailed.

Switch, general-use: A switch intended for use in general distribution and branch circuits. It is rated in amperes and is capable of interrupting its rated voltage.

Switch, general-use snap: A form of general-use switch so constructed that it can be installed in flush device boxes or on outlet covers, or otherwise used in conjunction with wiring systems recognized by the NE Code.

Switch, ac general-use snap: A form of general-use snap switch suitable only for use on alternating-current circuits and for controlling the following:

1. Resistive and inductive loads (including electric discharge lamps) not exceeding the ampere rating at the voltage involved.
2. Tungsten-filament lamp loads not exceeding the ampere rating at 120 V.
3. Motor loads not exceeding 80% of the ampere rating of the switches at the rated voltage.

Switch, ac-dc general use snap: A form of general-use snap switch suitable for use on either direct- or alternating-current circuits and for controlling the following:

1. Resistive loads not exceeding the ampere rating at the voltage involved.
2. Inductive loads not exceeding one half the ampere rating at the voltage involved, except that switches having a marked horsepower rating are suitable for controlling motors not exceeding the horsepower rating of the switch at the voltage involved.
3. Tungsten-filament lamp loads not exceeding the ampere rating at 125 V, when marked with the letter T.

Switch, isolating: A switch intended for isolating an electric circuit from the source of power. It has no interrupting rating and is intended to be operated only after the circuit has been opened by some other means.

Switch, motor-circuit: A switch, rated in horsepower, capable of interrupting the maximum operating overload current of a motor having the same horsepower rating as the switch at the rated voltage.

Switchboard: A large single panel, frame, or assembly of panels, having switches, overcurrent or other protective devices, buses, and usually instruments, mounted on the face or back or both. Switchboards are generally accessible from the rear as well as from the front and are not intended to be installed in cabinets.

Thermal cutout: An overcurrent protective device containing a heater element in addition to, and affecting, a renewable fusible member, which opens the circuit. It is not designed to interrupt short-circuit currents.

Thermally protected (as applied to motors): Refers to the words *thermally protected* appearing on the nameplate of a motor or motor-compressor and means that the motor is provided with a thermal protector.

Thermal protector (as applied to motors): A protective device that is assembled as an integral part of a motor or motor-compressor and that, when properly applied protects the motor against dangerous overheating due to overload and failure to start.

Trusses: Framed structural pieces consisting of triangles in a single plane for supporting loads over spans.

Utilization equipment: Equipment that utilizes electric energy for mechanical, chemical, heating, lighting, or other similar useful purposes.

Ventilated: Provided with a means to permit circulation of air sufficient to remove an excess of heat, fumes, or vapors.

Voltage (of a circuit): The greatest root-mean-square (effective) difference of potential between any two conductors of the circuit concerned.

Voltage to ground: In grounded circuits the voltage between the given conductor and that point or conductor of the circuit that is grounded; in ungrounded circuits, the greatest voltage between the given conductor and any other conductor of the circuit.

Watertight: Constructed so that moisture will not enter the enclosing case or housing.

Weatherproof: Constructed or protected so that exposure to the weather will not interfere with successful operation.

Web: Central portion of an I-beam.

Index

Page numbers in *italics* indicate illustrations. Page numbers followed by *t* indicate tables.

R

S

T

U

V

W

X

Z

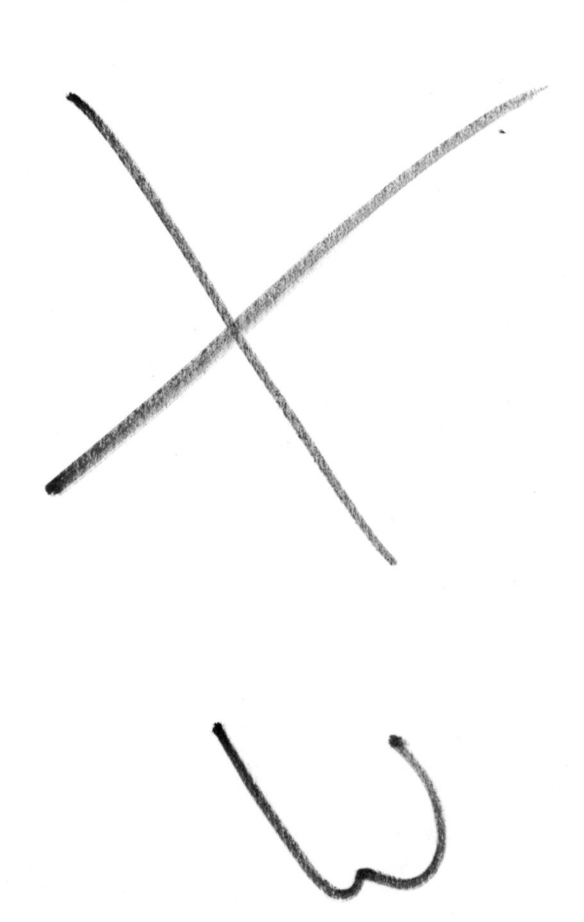